MASTER PONGO

Books in the Animalibus series share a fascination with the status and the role of animals in human life. Crossing the humanities and the social sciences to include work in history, anthropology, social and cultural geography, environmental studies, and literary and art criticism, these books ask what thinking about non-human animals can teach us about human cultures, about what it means to be human, and about how that meaning might shift across times and places.

OTHER TITLES IN THE SERIES:

Rachel Poliquin, *The Breathless Zoo: Taxidermy and the Cultures of Longing*

Joan B. Landes, Paula Young Lee, and Paul Youngquist, eds., *Gorgeous Beasts: Animal Bodies in Historical Perspective*

Liv Emma Thorsen, Karen A. Rader, and Adam Dodd, eds., *Animals on Display: The Creaturely in Museums, Zoos, and Natural History*

Ann-Janine Morey, *Picturing Dogs, Seeing Ourselves: Vintage American Photographs*

Mary Sanders Pollock, *Storytelling Apes: Primatology Narratives Past and Future*

Ingrid H. Tague, *Animal Companions: Pets and Social Change in Eighteenth-Century Britain*

Dick Blau and Nigel Rothfels, *Elephant House*

Marcus Baynes-Rock, *Among the Bone Eaters: Encounters with Hyenas in Harar*

Monica Mattfeld, *Becoming Centaur: Eighteenth-Century Masculinity and English Horsemanship*

Heather Swan, *Where Honeybees Thrive: Stories from the Field*

Karen Raber and Monica Mattfeld, eds., *Performing Animals: History, Agency, Theater*

J. Keri Cronin, *Art for Animals: Visual Culture and Animal Advocacy, 1870–1914*

Elizabeth Marshall Thomas, *The Hidden Life of Life: A Walk Through the Reaches of Time*

Elizabeth Young, *Pet Projects: Animal Fiction and Taxidermy in the Nineteenth-Century Archive*

MASTER Pongo

A GORILLA CONQUERS EUROPE

Mustafa Haikal

Translated by Thomas Dunlap

THE PENNSYLVANIA STATE UNIVERSITY PRESS
UNIVERSITY PARK, PENNSYLVANIA

Library of Congress Cataloging-in-Publication Data

Names: Haikal, Mustafa, 1958– author. | Dunlap, Thomas, 1959– translator.
Title: Master Pongo : a gorilla conquers Europe / Mustafa Haikal ; translated by Thomas Dunlap.
Other titles: Master Pongo. English | Animalibus.
Description: University Park, Pennsylvania : The Pennsylvania State University Press, [2020] | Series: Animalibus: of animals and cultures | Originally published in German in 2013 as Master Pongo : Ein Gorilla erobert Europa. | Includes bibliographical references and index.
Summary: "Relates the story of a juvenile gorilla named Pongo, brought to Europe in 1876 and housed at the Unter den Linden Aquarium in Berlin. Examines human-animal interactions and science at a time when the theory of evolution was first gaining ground"—Provided by publisher.
Identifiers: LCCN 2019055809 | ISBN 9780271082165 (cloth)
Subjects: LCSH: Master Pongo (Gorilla), 1874–1877. | Gorilla—Behavior. | Captive mammals—Germany—Berlin—History—19th century.
Classification: LCC QL737.P94 H356 2020 | DDC 599.884—dc23
LC record available at https://lccn.loc.gov/2019055809

Originally published in German as *Master Pongo: Ein Gorilla erobert Europa*

Printed in the United States of America
Published by The Pennsylvania State University Press,
University Park, PA 16802-1003

The Pennsylvania State University Press is a member of the Association of University Presses.

It is the policy of The Pennsylvania State University Press to use acid-free paper. Publications on uncoated stock satisfy the minimum requirements of American National Standard for Information Sciences—Permanence of Paper for Printed Library Material, ANSI Z39.48–1992.

Master Pongo is the most wonderful monkey I have ever seen, and I am a fellow of the Zoological Society and have visited all the living collections of natural history specimens in Europe and America.

—London correspondent
for the *New York Times*,
12 August 1877

CONTENTS

ACKNOWLEDGMENTS

From the very beginning, once I had conceived the idea for this book, I wanted to tell above all the story of Pongo and trust the power of the tale of his unusual life. I therefore held back with interpretations that would have interfered with the flow of the narrative, and my goal was never a work intended for a small circle of specialists. Still, I owe a debt of gratitude to many scholars, without whose help this book would never have seen the light of day. First and foremost I must mention Lothar Dittrich, Dietmar Stübler, Harro Strehlow, and Lothar Stein. I also received support in the form of material and advice from Lothar Diez, Marie-Eva Gamblin, Klaus Gille, Sabine Grunwald, Ingeborg Haikal, Saskia Jancke, Lisa Kaufmann, Ursula Klös, Frank Morgenstern, Bruno Schelhaas, Sibylle Stockmann, Julia Wicke, Christa Winkler, and especially Jan Hoyer. To all of them, and especially my wife, Marlies, a heartful "Thank you."

CHAPTER 1

"Gorilla Fever"

In the summer of 1876, Berlin was in the grip of "gorilla fever." Curiosity was already rising in the middle of June, as events were drawing closer and news reports were coming in, first from Liverpool and then from Hamburg. What the press was calling the "most gigantic ape known to zoology" was said to be on the way to the capital of the Reich and enjoying "perfectly good health." Interest seemed to be growing with every passing day. "Never and nowhere has a member of the animal kingdom been expected with greater eagerness than this gorilla, and never before has the final fate of an animal been the topic of such animated controversies and such astounding debates," reported the *Vossische Zeitung*. That does sound extraordinary, yet it created the wrong impression, for the great ape who was being announced, and who reached Berlin with a group of German Africa explorers, was a juvenile male less than thirty-two inches tall. His name: M'Pungu or Master Pongo, or simply Pongo.

"The gorilla in Berlin"—"the gorilla at the Aquarium": hardly an educated citizen of the city missed this event in the weeks that followed. Notwithstanding his youth and his scabby arms and legs, the ape was a sensation. At twenty thousand marks, he had cost nearly as much as a small villa, a fact that astonished contemporaries.

It made perfect sense, then, that the new arrival was put on display not in the Zoological Garden but in the Aquarium "Unter den Linden"—not at the outskirts of the city, but at its center, on Berlin's most magnificent and famous street. Henceforth the gorilla's headquarters would be here, close to the Brandenburg Gate. This would be the place where he held court and quickly charmed those flocking to see him. Being charmed was something very few of the visitors would likely have expected: too terrifying and frightening was the reputation that preceded this ape. For years, books and papers had presented readers with a monster—a monster with human features, no less. Even if there were more realistic views, unease about the "uncanny relative" was deeply rooted. The drawings in magazines continued to show a mythic, bear-like creature with massive canines, and the Brockhaus Encyclopedia was still describing the gorilla as one of "the most horrid creatures" imaginable.

Depiction of a gorilla, 1873

The Aquarium's new attraction, "the widely traveled anthropoid," created a different impression. Articles in the daily papers revealed astounding bits of information. The gorilla seemed not only curious and capable of

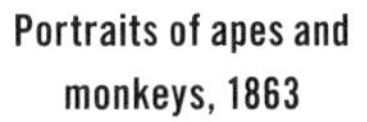

Portraits of apes and monkeys, 1863

learning, but also cheerful and amiable. Like an infant, he needed round-the-clock care, and he would not tolerate being left alone even for a minute. What did it matter, then, that the name M'Pungu was translated as "the devil," or that scientists believed the animal would reveal its "vicious nature" only when fully grown? At least for now, there was no trace of a bone-crushing monster, and before long the gorilla was considered "one of the most endearing and popular residents of the Reich capital."

The Aquarium, which also exhibited birds and mammals alongside fish and reptiles, had never seen such an onrush of visitors. The number of subscribers quickly doubled, and Berliners were coming in droves. The cage in which Pongo took up residence, close to the aviaries, was densely thronged with visitors crowded several rows deep. And when the director of the institute, the chemist Otto Hermes, took the gorilla to his official residence in the evening and stepped to the window one last time with him, Pongo caused a traffic jam in the street below. Eventually the police stepped in to restrain the curious onlookers and restore order outside the Aquarium.

Among those who took an interest in Pongo and announced their visits to Otto Hermes were many scientists. "The gorilla exerts an enormous attraction, and no other member of the animal kingdom has been subjected to such a thorough examination by the luminaries of our local experts in so brief a period after its arrival," wrote the *Vossische Zeitung*. The names we read are impressive. The advisory board in charge of the great ape's health alone was made up of three professors: the world-famous medical doctor Rudolf Virchow; the director of the veterinary school, Christian Gerlach; and the ethnologist Robert Hartmann. Together with an entire team of doctors they were responsible for Pongo's well-being while at the same time seeing him as an object of scientific study. For the experts, this was a unique opportunity: the most controversial creature that explorers and adventurers had ever discovered in the African jungles could now be closely observed on Unter den Linden.

CHAPTER 2

The Discovery of a Monster

FIRST REPORTS

The excitement surrounding Pongo and the frantic reports and press releases were chiefly driven by the belief that he was the first living gorilla successfully brought to Europe. While not true, this was asserted by nearly all sides. In 1876, the scientific description of gorillas went back a mere three decades. Many characteristics of the species remained unclear until Pongo's arrival, and even the stuffed specimens in the museums tended to baffle the scientists. Often treated with black oil paint and varnish, they frightened visitors with their gaping mouths and threatening gestures. Since no two specimens were alike, different pictures emerged. Similarly, nobody could answer the question of when seafarers and European traders had first encountered signs of the hominoid. Was the

Thomas S. Savage

Carthaginian Hanno really the first to observe gorillas during his voyage along the west coast of Africa around 470 B.C.E., or was it the Portuguese, who had controlled trade with that region since the fifteenth century? Accounts of giants and ape-like monsters, of cannibals and humans with tails, were legion, and for a long time it was taken for granted that tales about Africa blended experiences and hearsay, facts and rumors.

The English pirate Andrew Battel, who caused a stir with his account of "Angola and the adjoining regions," is no exception in this regard. His story, published as early as 1625, spoke of two monsters that were "very dangerous." According to Battel, "the greatest of these two monsters" was called "Pongo" by the natives, and while it had the build of a human, it was more like a giant than a man. The jungle-dwelling monster had deep-set eyes and hair "of a dunnish color," its face and hands being hairless. The author went on to relate this about the Pongos:

> They go many together, and kill many negroes that travel in the woods. Many times they fall upon the elephants, which come to feed where they be, and so beat them with their clubbed fists and pieces of wood that they will run roaring away from them.
>
> Those *Pongoes* are never taken alive, because they are so strong that ten men cannot hold one of them.[1]

So much for Andrew Battel, whose tale was reprinted many times and frequently cited. Other travel accounts with similar descriptions followed. As they did not shed much additional light, it was not until 1847 that the clergyman Thomas S. Savage attracted attention with new information. An experienced missionary, the American had spent more than ten years on the west coast of Africa. His first two wives had died there of tropical diseases, and the grueling climate had taken its toll on him as well. Worn out and in poor health, he set out on his journey home, stopping near the mouth of the Gabon in April 1847 to visit his colleague John Leighton Wilson. Savage was not only a missionary but also a physician and naturalist. He studied insects at every opportunity and had even written an article about chimpanzees. That would now prove an advantage, for his host, a fancier of African curiosities, showed him a skull shortly upon his arrival. Unusually large and sporting a pronounced cranial crest, it could not be assigned to any known

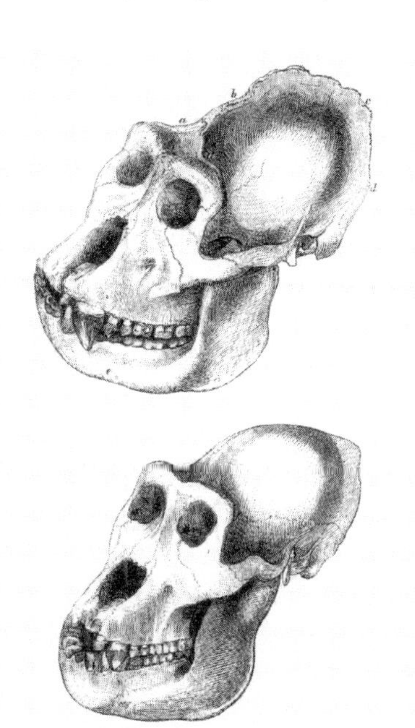

The drawings sent by Savage to Europe and America in 1847

species. The American immediately understood what this meant, and in the weeks that followed he made every effort to obtain more information. By the time he departed in early May, he had procured additional skulls and bones and had carefully recorded the tales of indigenous hunters.

Two English captains also had their interest piqued. While Savage was still sailing across the ocean, they made further inquiries. Back home good money could be made with specimens of African animals, especially if the species were still unknown. And so, without the American having an inkling of it, a race was under way, and gorilla skulls were being brought to Europe as well. Savage himself, who was a little naïve in such matters, wrote to two English specialists and reported his discovery with the help of drawings. Luck was on the side of the missionary, as he eventually beat out all his rivals and became the first to publish the discovery. His article, which appeared in Boston in 1847, had two parts: a brief description of the gorilla and everything known about it, and a study—nearly twice as long—of the bones he had found.[2] For the second part of the article Savage had consulted an expert, Jeffries Wyman, who taught anatomy at Harvard College. Together they summarized what they knew and had ascertained from their examination of the skulls and bones. Though Savage had never set eyes upon a living gorilla, he drew a very vivid picture. He dismissed the old stories—which he deemed "silly"—of a monster that carried off native women, along with tales of gorillas vanquishing even elephants. Instead of presenting a spine-tingling tale, he offered a kind of "WANTED" poster, with information about hair color, size, diet, and the social life of the animals. But Savage, too, agreed that these animals were

> exceedingly ferocious, and always offensive in their habits, never running from man as does the Chimpanzée. They are objects of terror to the natives, and are never encountered

> by them except on the defensive. The few that have been captured were killed by elephant hunters and native traders as they came suddenly upon them while passing through the forests.
>
> It is said that when the male is first seen he gives a terrific yell that resounds far and wide through the forest, something like *kh—ah! kh—ah!* prolonged and shrill.

In the end it was Wyman, not Savage, who suggested the name that would soon be on everyone's lips. Following a Greek source, he called the anthropoid "gorilla."[3] The term comes from a report by the previously mentioned seafarer Hanno, and it triggered debates among the experts. To this day it is unclear whether Hanno did in fact observe gorillas on his voyage along the west coast of Africa. His description could well have been of chimpanzees or Pygmies.

BONES AND CADAVERS

The name "gorilla" spread as quickly as the article by the two Americans. At the beginning of the nineteenth century, several naturalists had proclaimed that the discovery of new species of large animals was highly unlikely. Now the article from the other side of the Atlantic proved them wrong. This called for new questions and new answers. Ever since the beginning of the modern sciences and the Age of Enlightenment, scholars in the European metropolises had been in search of precise knowledge. More than ever before, they sought to organize the phenomena of nature into scientific systems, and they studied animals, plants, bones, and fossils with curiosity and meticulous care. Natural history collections gave rise to public museums of natural history, and disciplines such as biology and zoology emerged at the universities out of the old natural history.

Richard Owen

One of the men to whom Savage had sent drawings of the gorilla skulls from Africa was the Englishman Richard Owen. A zoologist and paleontologist, known today largely for coining the term "dinosaur," Owen also studied anthropoids. Among the most influential men in the world of Victorian science, he was forced to concede that the honor of the first description of the gorilla belonged to two outsiders. Still, he did not forgo an examination of the "new" skulls that arrived from Africa in December 1847. Only two months later he presented his finding to the Zoological Society of London. It would not be Owen's last contribution to the gorilla question. As an anatomist, he compared the most diverse animal species, and in highly regarded studies he returned repeatedly to the anthropoids.

For now, however, everything depended on when the next boats with new trophies would reach Europe. To make progress in identifying and describing the gorilla, the experts needed additional "material"—not only skulls and skeletons but living specimens, if possible. Since the latter seemed too much to hope for, they made do with specimens preserved in alcohol. And so the years that followed witnessed a series of unusual transports.

Eerie scenes transpired when the barrels containing the dead anthropoids were opened after their long voyage across the sea. The first episode took place in Paris around 1852. In the rivalry among the scientists, fortune favored the experts at the renowned Muséum national d'Histoire naturelle: the next discoveries would be made along the Seine, not along the Thames.

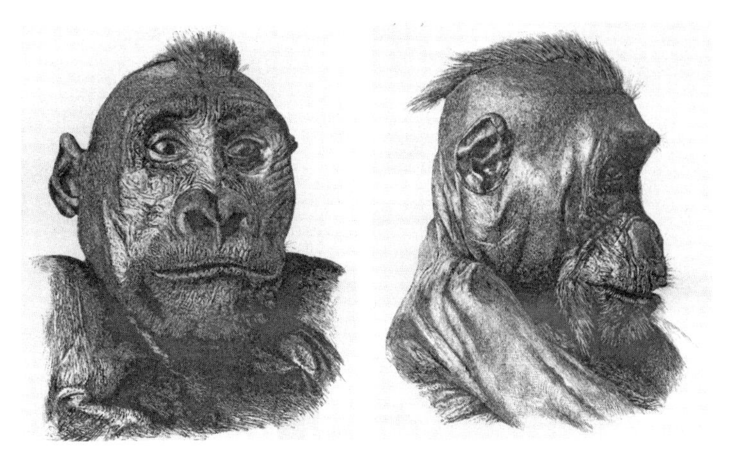

Head of a gorilla preserved in alcohol

Tense excitement was in the air when the valuable "gorilla cargo" from a warship was handed over to French scientists in January 1852. It became clear very quickly that at least the larger of the two specimens was showing signs of considerable decomposition. As with many transports of this kind, the alcohol had done a poor job of conserving the bodies. Upon opening the barrels, the scientists encountered clumps of hair that had become detached from the skin. The stench left even hard-boiled experts breathless, and

Illustrations from *Archives du Muséum d'Histoire Naturelle* (Paris, 1858)

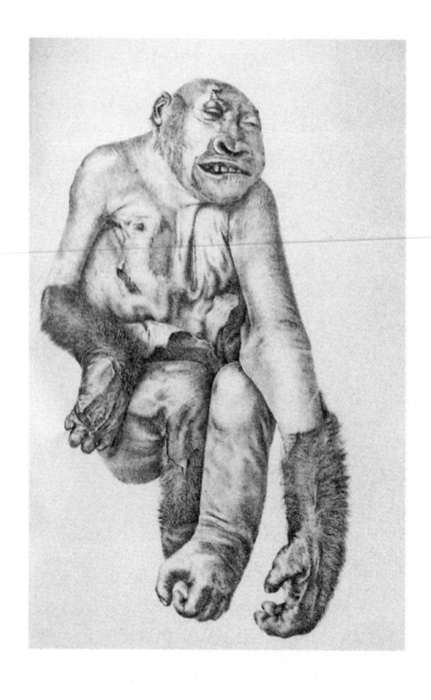

as we read in a report about a similar event in London, it would have sufficed to "poison an entire regiment of soldiers." There was a more serious problem, though: the deformations produced by the decay conveyed a distorted picture.

For example, the rubber-like stretching of the lips caused the teeth to be exposed in the most frightening manner. The unnaturally bloated belly and the mostly brownish color of the fur were also the result of putrefaction. But the scientists around the zoologist Isidore Geoffroy Saint-Hilaire were not deterred by any of this. The Frenchman, who was the first to realize that the gorilla was a distinct species, arranged for the careful processing of the cadavers. Following their arrival, the dead bodies were photographed and drawn, skinned and dissected. Two *moulages* (wax masks) were intended to capture their facial features with the utmost fidelity and preserve them for posterity. It would be years, though, before all the findings of the examinations were available, and it was not until 1858 that Geoffroy Saint-Hilaire wrapped up all the work.

The illustrations, above all, caused a stir well beyond Paris. For example, in May 1855 the *Illustrirte Zeitung*, one of the major family magazines in Germany, published three lithographs that allowed readers to gaze upon the deformed gorillas. The impression became even more dramatic when the museum's draftsman, the artist and zoologist Marie-Firmin Boucourt, created a color portrait depicting the gorilla standing upright next to a tree with its mouth wide

Illustration from *Archives du Muséum d'Histoire Naturelle* (Paris, 1858)

open. In this posture it was reminiscent of the Wild Man, a terrifying figure of myth and fable.

The specimens and moulages of the anthropoids also drew attention. Put on display at the Paris World's Fair in 1855, they were subsequently brought to the Natural History Museum of the Jardin des Plantes, which is where the sculptor Emmanuel Frémiet studied them intensely. When he exhibited the result of his work in the French capital in 1859, it caused a scandal. The sculpture, titled *Gorilla Carrying off a Woman,* a brutal and savage scene full of drama, touched on sexual taboos and reinforced existing fears.[4] The "sinister ape from Africa's forests" was now causing a stir in the heart of Europe: the "monster" had been given a face.

Gorilla sculpture, E. Frémiet, 1887

THE "MONKEY THEORY"

The same year that Frémiet exhibited his gorilla sculpture in Paris, an epochal book was published in London. It came from the pen of the naturalist and world traveler Charles Darwin and represented the fruit of years of studies. Titled *On the Origin of Species by Means of Natural Selection, or The Preservation of Favoured Races in the Struggle for Life,* its German translation was available only a few short months later. On the surface, the lengthy work seems to have little to do with the gorilla. There is not a single mention of the name of the ape, and other apes and monkeys are mentioned only in passing. But the first impression is deceiving. Even if the Englishman preferred to write about pigeons and dogs, his theories raised questions—questions that had already been debated for some time and would now rock the traditional picture of the world. For Darwin was providing nothing less than a key to the

Charles Darwin

history of life, to the processes of evolution. All living things, his theory asserted, had developed from common ancestors. It was not divine will that had given rise to them, but chance and natural selection. And why—so the almost ineluctable conclusion—should the processes that held for wild and domesticated animals not apply also to human beings? And if human beings, with all their innate intelligence, had emerged from the animal world, who were their ancestors and animal relatives?

Even though Darwin shied away from this debate and avoided any public speculation, he was unable to stave off the controversy. Opposing camps of opponents and supporters of his theory formed quickly, and the conflict intensified. Richard Owen, too, spelled out his position in the summer of 1860. His attempt to prove the unique nature of humankind by locating a brain structure found only in humans, the hippocampus minor, ended in defeat. The criticism of Owen was devastating. A new generation of scientists, headed by the biologist Thomas Henry Huxley, refuted his claims and set out to destroy his reputation. Although Darwin himself participated in the battles only from the background, privately he recorded his opinion in a notebook. In a diagram of evolution that he sketched in brown ink for his own understanding in April 1868, he identified the gorilla and the chimpanzee as humanity's closest relatives.[5]

In the end it was the gorilla that captured the interest of the public. It was linked to Darwin's theory like no other animal. That had to do not only with the dismal and vague picture that existed

Paul Du Chaillu

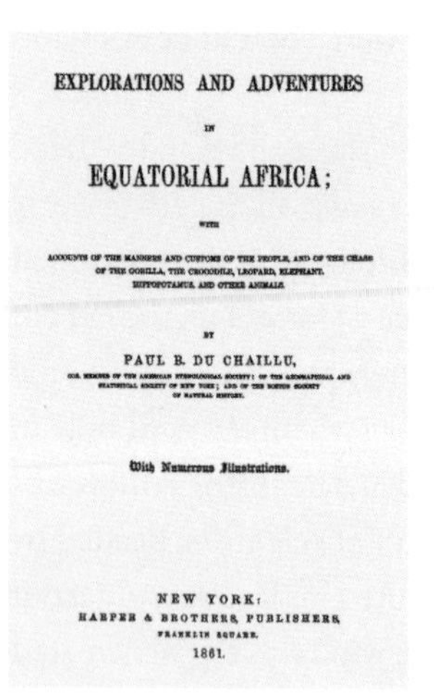

EXPLORATIONS AND ADVENTURES

IN

EQUATORIAL AFRICA;

WITH

ACCOUNTS OF THE MANNERS AND CUSTOMS OF THE PEOPLE, AND OF THE CHASE OF THE GORILLA, THE CROCODILE, LEOPARD, ELEPHANT, HIPPOPOTAMUS, AND OTHER ANIMALS.

BY

PAUL B. DU CHAILLU,

COR. MEMBER OF THE AMERICAN ETHNOLOGICAL SOCIETY; OF THE GEOGRAPHICAL AND STATISTICAL SOCIETY OF NEW YORK; AND OF THE BOSTON SOCIETY OF NATURAL HISTORY.

With Numerous Illustrations.

NEW YORK:
HARPER & BROTHERS, PUBLISHERS,
FRANKLIN SQUARE.
1861.

of it; chance also played a role. The same publishers who issued the German translation of *On the Origins* in 1859 released a sensational volume two years later: the travelogue of the adventurer and zoologist Paul Belloni Du Chaillu. A Franco-American whose exact date of birth is still disputed, Du Chaillu had lived on several continents—including Africa—as a child and young man. As an American citizen he traveled from New York to Gabon in 1856 with support from the Academy of Natural Sciences of Philadelphia. For three and a half years he traversed that country as its first explorer, in the process penetrating up to five hundred kilometers into the interior. His account, *Explorations and Adventures in Equatorial Africa,* became a great success and was translated into numerous languages. It is a story of adventure and daring, at the heart of which is the hunt for gorillas. "Suddenly," writes Du Chaillu,

> the woods were at once filled with the tremendous barking roar of the gorilla.
>
> Then the underbrush swayed rapidly just ahead, and presently before us stood an immense male gorilla. He had gone through the jungle on his all-fours; but when he saw our party he erected himself and looked us boldly in the face. He

Gorillas; below, scene after the account by Paul Du Chaillu, 1863

Du Chaillu,
Boston, ca. 1869

stood about a dozen yards from us, and was a sight I think never to forget. Nearly six feet high (he proved two inches shorter), with immense body, huge chest, and great muscular arms, with fiercely-glaring large deep gray eyes, and a hellish expression of face, which seemed to me like some nightmare vision: thus stood before us this king of the African forests.

He was not afraid of us. He stood there, and beat his breast with his huge fists till it resounded like an immense bass-drum, which is their mode of offering defiance; meantime giving vent to roar after roar.

The roar of the gorilla is the most singular and awful noise heard in these African woods. It begins with a sharp *bark*, like an angry dog, then glides into a deep bass *roll*, which literally and closely resembles the roll of distant thunder along the sky. . . .

. . . And now truly he reminded me of nothing but some hellish dream creature—a being of that hideous order, half man half beast.[6]

Skillfully exploiting his adventure, Du Chaillu gave many lectures in England and the United States. Additional travels followed; new books were published. In this way, Du Chaillu shaped the picture of the gorilla that was held by large segments of the public. But other experts, finding some of the things recounted by the gorilla hunter too bombastic and audacious, mistrusted his accounts, some going so far as to consider them invented. Although he had furs and skeletons of the ape to show, that alone was not enough for the critics. For example, even several years later the German zoologist Alfred Brehm was still describing Du Chaillu's tales as "a wondrous mix of truth and fiction," though this certainly did not keep him from citing them again and again.

When the first volume of the animal encyclopedia *Brehms Tierleben* (*Brehm's Life of Animals*) was published in 1864 with Du Chaillu's description, Darwin's name and the so-called monkey theory were already on everyone's lips. "Monkey or not!—Father? Uncle? Cousin? Out with it!"—ran a title in *Kladderadatsch*, a German satirical magazine. Behind this jaunty call for clarity stood a simple thesis: since Darwin's theory remained beyond the comprehension of many of his contemporaries, they reduced his voluminous book to the assertion that man was descended from the apes. Even though neither Darwin himself nor his followers ever claimed as much and merely posited a common ancestor, this idea proved tenacious. It was with very different eyes that the public now read tales about the gorilla and listened to the stories of the Africa

Caricature, *Punch*, 1861

Charles Darwin, caricature, 1871

explorers. That this giant ape, this "black monster" from the African jungle, was supposedly a relative of man, however distant, created a sense of unease. After all, of all creatures, the gorilla seemed the least suited for such a role. And in an effort to dispel thoughts that were deeply upsetting, the topic was given a humorous twist. In satirical magazines and colorful periodicals, in dailies and annual calendars, the gorilla enjoyed a very different kind of career as a tragico-comical monster, becoming after the middle of the century the most frequently sketched ape. Darwin himself, who spoke in detail about the topic for the first time in 1871 in his book *The Descent of Man, and Selection in Relation to Sex,* was "honored" in the same way: depicted as half ape, half man, he served as the model in the search for the missing link. Far from being irritated by that kind of attention, he collected the many caricatures and kept them in a cardboard box.

JENNY AND MAFOKA

Notwithstanding its career as a figure of art and imagination, the gorilla remained a challenge. In countless works, scientists and laypeople speculated about a wealth of questions. What good were new eyewitness accounts if they merely tended to reinforce the chaos of opinions? Yet a living specimen, a real-life gorilla, was nowhere in sight. To be sure, there were plenty of attempts to bring juvenile animals captured by natives to Europe. Thomas S. Savage had told of a Frenchman who was not deterred by the difficulties this entailed. However, the gorilla he had obtained died during the crossing.

Additional reports can be found in scientific journals and travelogues, and we know of the fate of about a dozen animals up to 1876. These are heart-rending and moving accounts, which seem inexplicable in the frequency of accidents they relate: the animals died of diarrhea, came down with scurvy and other illnesses, fell overboard during the crossing, and starved because the crew ran out of bananas. Around 1861, Richard Owen related that a gorilla brought along by a captain died just as the ship was entering the port of Le Havre.

A. D. Bartlett, ca. 1858

Once, however, the attempt succeeded—but then failed in the end after all. In the late summer of 1855, a young animal was brought from the Congo to Liverpool and sold to the owner of Wombwell's Traveling Menagerie. The widow of the exhibitor George Wombwell believed that the new acquisition was a chimpanzee, and no one seems to have disputed her. Dressed in girl's clothing and given the name Jenny, the female gorilla remained alive for about seven months and could be seen in a number of towns and cities in northern England. And that was how Charles Waterton, a famous world traveler and pioneering conservationist, had a chance to observe Jenny. His reports about these encounters were printed in a local paper and published in 1857 in a volume of essays. Even a cursory look at the extensive text is enough to dispel all doubts:

> Jenny has no appearance whatever of a tail, for she is a veritable ape. Her skin is as black as a sloe in the hedge, whilst her fur appears curly and brown. Her eyes are beautiful; but there is no white in them; and her ears are as small in proportion, as those of a negress.
>
> Whilst apes in general, saving one, have little more than two apertures by way of nose—Jenny has a large protuberance there. It is flattened; and one might suppose, that some officious midwife had pressed it down with her finger and thumb, at the hour of Jenny's birth.[7]

A second piece of evidence is just as unequivocal. The owner of the menagerie had a photo taken of Jenny. Several years after the death of the ape, the picture came into the possession of Abraham Dee Bartlett, the superintendent of the London Zoo. He was the first to recognize that he was looking at a gorilla, and he gave the image to one of England's best animal illustrators. Josef Wolf, German by birth, used the photograph to create a chalk drawing that showed Jenny sitting on a forked branch surrounded by tropical plants.[8]

Jenny, chalk drawing

These tales were simply new fodder for the public's "gorilla yearning." Every bit of news was eagerly received, and at times the public allowed itself to be misled. The specimen described in the *Liverpool Mercury* in 1862 was in reality a chimpanzee, and the gorilla displayed by the crooked owner of a traveling menagerie in Berlin turned out to be a black baboon. But the greatest excitement was triggered by a female chimpanzee that lived in the Dresden Zoo from July 1873 to December 1875. Experts came to very different conclusions about this animal, by all accounts a particularly wild specimen. While some highlighted its dark facial color and discovered further "gorilla traits," others challenged this identification. The discussion soon took on an implacable tone. The experts accused one another of "a disregard of all facts" and "petty grumbling" and demanded "complete clarification" and "strict objectivity." Mafoka—the name given to the ape—was captured in portraits, presented to the scientists in private audiences, and examined. The large family magazines also discovered the story and sent their own reporters and illustrators.

Portrait of Mofaka, 1875

Alwin Schoepf, the director of the Dresden Zoo who cared for Mafoka, was unperturbed by the constant stream of visitors. He knew that the publicity was good for business and that the controversy was therefore useful. A substantial portion of the zoo's revenue was courtesy of this unusual animal. Schoepf therefore kept a close eye on his charge's health, and he was becoming increasingly concerned. When Mafoka died from tuberculosis during the night of 14/15 December 1875, that event, too, made the headlines. One more time, we read in one journalist's account, the ape "wrapped its hands around the director's neck" before sinking back down onto its deathbed "weak and limp." Two days later, the *Dresdner Nachrichten* published a lengthy and heartfelt obituary: "Now the ape has died at the height of its importance, courted by the leading men of science, ardently desired by the zoological gardens,

Taxidermied skin of Mofaka (center) in the Königliches Zoologisches Museum Dresden, ca. 1877

praised by the stockholders as a pearl of this garden, coddled by the administration, and cared for by the keepers with distinction and special devotion."

The dispute itself was resolved only a few months later. For just as the experts had quarreled over the identification of the ape, they now fought over the corpse. The very night of Mafoka's death, efforts were made to reserve the mortal remains for the capital's Anatomical Museum, where "all possible preparations have been made to take photographs, prepare color drawings, make plaster casts, and methodically dissect the body parts." But Adolf Bernhard Meyer, the director of the Royal Zoological Museum in Dresden, had acquired the valuable cadaver for the sum of four hundred talers and wouldn't dream of giving it up. Even though he was obligated to do just that by a decision of the zoo's administrative board, he delayed sending it to Berlin to allow his own institution to conduct its research. In the end it was Pongo's arrival that took the air out of the controversy in the summer of 1876. It instantly became clear that some observers had ventured too far out on a limb in their identification of the "Dresden ape." Only now, with the second gorilla to reach Europe alive, were scientists able to settle the old case. At the Aquarium "Unter den Linden," and only there, could one observe a true gorilla. Over the next eighteen months, the animal would attract widespread interest, prompt the experts in Berlin, Hamburg, and London to pen more precise descriptions, and move the public more profoundly than any ape had ever done before.

CHAPTER 3

The Research Station on the Coast of Loango

A COLLECTION OF PHOTOGRAPHS

The next path on the trail of the gorilla leads us to Leipzig. The images we are looking for are in an archive at the outskirts of the city. They are the photographs of the German Loango expedition, published in August 1876. Only a few of the albums still exist worldwide, and one is in the possession of the Leibniz-Institut für Länderkunde (Leibniz Institute for Regional Geography). The display box contains seventy-two photographs, arranged on thirty-five sheets of yellow-gray cardboard. And there he is, on page 22: the gorilla. The photograph of the small ape is barely larger than a pack of cigarettes and so touching that no one can escape its emotional power. Pongo is resting against a wooden post, his head leaning to one side and his eyes closed. To the left and right are four portraits of inhabitants of the coastal region: “The old man of the mountain, ruler in the village of Luzalla”—“The priest of the earth spirit N’Kissi”—“Cabinda Negro”—“Young Babongo,” read the captions. But they are merely extras, for the most important image is that in the center; everything revolves around it.[1]

The album, with its sumptuous ornamentation, is not the only reason for my visit to the nondescript modern building. Another collection can be found in an archival folder from the papers of the geographer Eduard Pechuël-Loesche. It contains dozens of animal portraits, eleven of which show Pongo. No fewer than five images were taken in Africa in October 1875. The still enfeebled gorilla is sitting in a corner, lying on a blanket, hunched up against a wall—asleep or semiconscious. These are followed by six images from the time at the Aquarium. It is a very different ape that we encounter in these plates. Larger and stronger now, he is looking into the camera. In other ways, too, the papers contain many surprises. Eight boxes contain statistics and excerpts, notes and articles, letters and newspaper clippings. Linguistic material concerning the Bantu language,

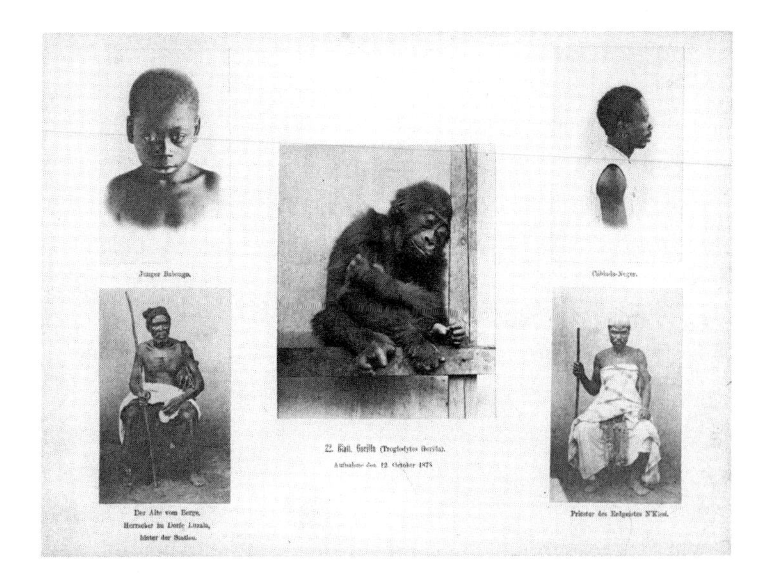

"Notes on the music of the primitive races," observations about meteorology—all of this follows. And time and again, the manuscripts concern the coast of Loango, a landscape that is today part of Angola. Pechuël-Loesche spent two years of his life there, at the German research station Chinchoxo (see illustration on page 26). His evocative diary from that time runs hundreds of pages and provides more exciting reading than any novel. But the story of the station and of the Africa expedition associated with it begins earlier than the gorilla, and it all started in Berlin.

AUDACIOUS IDEAS

Although the capital of the German Empire still struck many observers as provincial in 1872, it was a metropolis on the rise. Wherever one looked there was new construction and investment, new stores opening their doors. It was hard for anyone to escape the city's energy, and the members of the Gesellschaft für Erdkunde zu Berlin (Berlin Geographical Society), too, were pursuing far-reaching plans. Following the creation of the empire, they were eager to finally secure for the fatherland the place to which England and France had long since laid claim. At stake was nothing less than the mapping of the world, the discovery of the last unexplored regions. Africa, above all, had captured their imagination. Though the "dark continent" was closer to Europe than any other destination, its interior was the least known. The explorations of David Livingstone and Henry Stanley had drawn eager attention in Germany, and their routes had been traced out on maps. In 1872, Georg Schweinfurth arrived in Berlin and reported on his three-year journey to the Pygmies and other tribes. His return provided a boost to the ongoing endeavors. This much seemed clear: whoever failed to organize his forces quickly and resolutely would be too late in the race among the nations to uncover the last secrets and exploit Africa, and would have to be content to watch from the sidelines.

This situation had been troubling the physician and naturalist Adolf Bastian for some time. Around the middle of the century he had traveled to Australia and India, and in 1857 he had visited Angola. Fifteen years later he was among the founders of German ethnology, and his name was appended to numerous publications. Now, as the chairman of the Berlin Geographical Society, he took the initiative and invited representatives of the country's geographical societies to Berlin for deliberations. After extensive preliminary work, in which delegates from Leipzig, Hamburg, Munich, and other cities participated, the Deutsche Gesellschaft zur Erforschung Äquatorial-Afrikas (German Society for the Exploration of Equatorial Africa) was established in April 1873. It took up residence in Berlin's Kronenstraße, where equipment for the planned expedition soon filled several rooms.[2]

Adolf Bastian

Since the Society published a bulletin (*Korrespondenzblatt*), its activities are well documented. As always—in this regard little had changed in the empire—the initial concern was money. Contributions to the expedition were falling well short of expectations. While the urgent pleas and national appeals received a sympathetic hearing, no one, with the exception of one bank director, could be prevailed upon to make a substantial donation. The Reichstag, too, refused to authorize ongoing support; only the emperor granted a

Illustrirte Zeitung.

Nr. 1614.] Erscheint jeden Sonnabend. Leipzig, 6. Juni 1874. Vierteljährlicher Abonnementspreis 2 Thlr. [LXII. Band.

Die deutsche Expedition zur Erforschung Innerafrikas.

Ein Jahr ist jetzt verflossen, seit durch einmüthiges Zusammengehen der geographischen Gesellschaften Deutschlands die Afrikanische Gesellschaft zu Berlin begründet wurde, deren Zweck die wissenschaftliche Erschließung der noch unbekannten Gebiete Centralafrikas ist.*) In der That konnte gegenwärtig keine bedeutungsvollere Expedition geplant werden als eine von der Westküste Afrikas, von den nördlichen Kongoländern nach Osten hin aufbrechende, die nach den Gebieten der großen, durch Livingstone entdeckten Seen vordringt und dort die mit einem Schlag die Räthsel der Gestaltung Innerafrikas löst. Auch die Engländer sind in dieser Richtung bereits thätig, auch sie haben erkannt, daß das wichtigste, für unsere Zeit gleichsam noch aufgesparte Forschungsgebiet zu beiden Seiten des Aequators in Afrika liegt. Dank der Deutschen Gesellschaft zur Erforschung Aequatorialafrikas sind wir jedoch auch im Stande, bei der Lösung dieser großen Aufgabe maßgebend mitzuwirken, und der Beginn, die Einleitung des ganzen Unternehmens ist so vielversprechend, daß wir wol mit einiger Genugthuung auf das innerhalb eines Jahrs Geleistete zurückschauen dürfen.

Versuchen wir es, einen kurzen Ueberblick des bisherigen Verlaufs der Expedition und der von ihr erreichten Resultate zu geben. Die erste Nachricht, welche Europa erreichte, war keine erfreuliche; sie meldete uns den Schiffbruch des Dampfers Nigritia auf dem Carpenter-Felsen an der Küste von Sierra Leone am 14. Juni 1873. Fast alle Instrumente, die ganze Ausrüstung gingen verloren, und Dr. Güßfeldt und v. Hattorf retteten nur das nackte Leben. Aber trotz des so bedeutenden Verlustes, den sie erlitten, trotz der traurigen Lage,

*) Vergleiche darüber und über die ersten Schritte der Expedition den Artikel von Gerhard Rohlfs in Nr. 1564 der „Illustrirten Zeitung" vom 21. Juni 1873.

subsidy, though it had to be applied for anew every year.[3] And so it was smaller amounts that trickled in, but these did add up, and for the moment they filled the coffers. Fortunately, one member of the expedition, Paul Güßfeldt, made six thousand talers from his personal wealth available to pay for the cost of his own equipment. Enough money, then, to launch the expedition in May 1873, but too little to sustain it permanently.

It was an audacious plan that the board of the society was pursuing with the expedition. This time, in contrast to all previous attempts, the continent was to be conquered from the southwestern coast. The hope was that this advance into the interior, the exploration of the Congo and its tributaries, would attract worldwide attention. More still: the initiators of this project had even made it their mission to traverse Africa all the way to the Indian Ocean. Ambitious goals, then, that called for great strategic vision. The plan was to send the men "into the field" from three sides, in addition to setting up a research and base station. Bastian himself traveled to the coast of Loango that very June on a steamer of the Portuguese

Africa line. His letters to the board sounded highly promising and contained a wealth of gripping details. That they also included reports about gorillas was received with enormous interest in Berlin. For example, Bastian reported on a European trader who had already raised young animals three times in the region around "Chicambo," which is why there was no better place "to conduct observations about these forest dwellers [*Waldmenschen*]."[4]

ON THE WAY TO AFRICA

Bastian's fifteen-week sojourn was merely a warm-up. The actual team of the expedition was made up of other men. Nearly fifty applications had been received in Berlin as soon as word of the Africa plans got out. Submissions from all over Europe and even the United States had to be reviewed and responded to. Interest in joining the expedition was thus considerable, but not everyone who stepped forward fulfilled the necessary requirements. There is good reason why a journey into the interior of Africa was still considered an adventure. Tropical diseases especially—including yellow fever and malaria, against which explorers tried to protect themselves with quinine and low doses of arsenic—claimed many lives. Participants had to be young and resilient, but at the same time sufficiently qualified. A university degree or solid professional training was as much expected as a sharp eye, powers of observation, and strength of will. Still, many dreamed of this chance for its promised fame and better opportunities for advancement. Anyone who survived, and in

P. Güßfeldt

the nineteenth century around two-thirds of explorers did, would later find his name in a lexicon.

The expedition that set out in stages beginning in 1873 consisted of at least eight members. Its leader, Paul Güßfeldt, had studied mathematics and natural sciences and had received his habilitation (a post-doctoral university qualification) in Bonn. A passionate alpinist whose first ascents were attracting attention throughout the country, he had still been climbing in the mountains of the Mont Blanc massif a short time before.[5] The mariner and geographer Eduard Pechuël-Loesche, meanwhile, was familiar not only with the coastal regions of both Americas but also with the island world of the Atlantic and Pacific Oceans. Several members held doctoral degrees, and three later became professors. For now, though, the scientists were at the beginning of their careers, and they set out for Africa filled with hope. Almost always, the journey began in Hamburg and led via London to Liverpool, the port that had been a major staging ground for the trade with slaves from Africa in the eighteenth century. It was now the starting point for the ships of the British and African Steam Ship Company, on which most of the travelers had booked passage: in pairs or alone, on different steamships, and with precise instructions in their extensive baggage.

In August 1873, two additional members of the expedition were named: the military doctor Julius Falkenstein and Otto Lindner, a young mechanic. This was not the first time that Falkenstein had felt the pull of far-off lands. Six years earlier, he had applied for the German North Pole expedition, but he had been too late. Still unmarried and stationed in the capital for some time now, the native Berliner tried again. The thirty-one-year-old Falkenstein, who would care for the gorilla Pongo over the course of the journey and bring him to Europe, hailed from a family of doctors. His father, the son of a Jewish merchant from Prussian Friedland, had converted to Protestantism in 1830 and had practiced as a physician in the Charité. Julius Falkenstein continued the profession. After graduating from the French Gymnasium, where he learned a lot about the ancient Greeks and little about the natural sciences, he began a course of study at the Pépinière. This famous training center for Prussian military doctors, the Medicinisch-chirurgisches Friedrich-Wilhelm-Institut, was considered praxis-oriented and possessed a decisive advantage: the education and all exams were free. This allowed members from less affluent families to study here, with famous examples ranging from Rudolf Virchow to Gottfried Benn.

J. Falkenstein

Even before receiving his medical license, Falkenstein went to war and served in the campaign against Austria in the summer of 1866. Only afterward was he able to complete his studies and obtain his doctorate with a dissertation in Latin about the treatment of typhus. We do not know whether the young doctor was genuinely enthusiastic about medicine. What is clear is that he also attended lectures on zoology during his training and joined the German Society of Ornithology. His name appears alongside that of Alfred Brehm and Otto Hermes in the minutes of the society, which met at the end of 1872 in the brewery "Unter den Linden." At that time Falkenstein was serving on the staff of General Gustav von Lauer, the personal physician to the emperor. Von Lauer promoted the career of his promising staffer and made it possible for him to take part in the Loango expedition, for which he was granted a leave of absence from the army.

On 20 September 1873, the doctor and the mechanic under his command set out for Africa with fourteen chests soldered shut and two shaggy black German shepherds. The *Benin*, the steamship

Landing dock in the harbor of Liverpool, ca. 1858

Wreck of the steamer in which Paul Güßfeldt
was shipwrecked off the coast of Africa in 1873

that would take them from Liverpool via Madeira and the Canary Islands to the west coast of Africa, was fully loaded. A small number of passengers shared the space with a great deal of cargo. Because of the frequent stops in ports along the way, the crossing took thirty-six days, enough time to organize one's thoughts and keep a journal. For Julius Falkenstein, a dream was becoming reality. Since childhood he had fantasized about such a trip, had conjured in his mind the sea and exotic countries. The passage itself was not without risks. Illnesses occurred, and on the ships they encountered heading the other way (on which the doctor treated several patients), he witnessed his first burials at sea. Seven Europeans had died on the *Ambria* alone, which was returning from the Congo. And there was another danger: twice the *Benin* ran aground off the coast, though she always succeeded in freeing herself after a brief period. Paul Güßfeldt, the leader of the expedition, had suffered a very different fate: during a major maritime disaster on 14 June 1873, he had to watch as most of his valuable equipment sank into the brackish seawater.

CHINCHOXO

Julius Falkenstein and Otto Lindner were headed for Chinchoxo, a trading station founded by the Portuguese about fifty-five miles north of the mouth of the Congo. With the help of local workers, Paul Güßfeldt had been able, over the preceding months, to spruce up the abandoned and dilapidated houses and to put down wooden floorboards. The site of the station, about seventy by fifty-seven meters in size, was located on a grassy plateau directly on the Atlantic. However, the images and drawings that exist of the place do not convey a clear picture. The members of the expedition did not by any means live in the forlorn isolation suggested by the images. A Dutch outpost was a mere eight hundred meters away, and the village of Lusala about a kilometer. In fact, the Germans were in every respect dependent on their neighbors. Without the help of the European traders and the local population, they would not have been able to erect and maintain their station. The coast of Loango had long since been within the spheres of the European trading companies, the advance posts of colonization. The region was covered by a network of outposts, villages, missionary stations, and smaller settlements. Anyone looking for "untouched Africa" here was centuries too late. Still, Falkenstein's first letters sounded hopeful. He described the "fine sight of the research station located on a mountain ridge" and was sorry he could not yet greet the passing ships with the German flag. The account of his arrival in Chinchoxo in his later book makes for very different reading. With hindsight about the outcome of the expedition, he described the situation more cautiously:

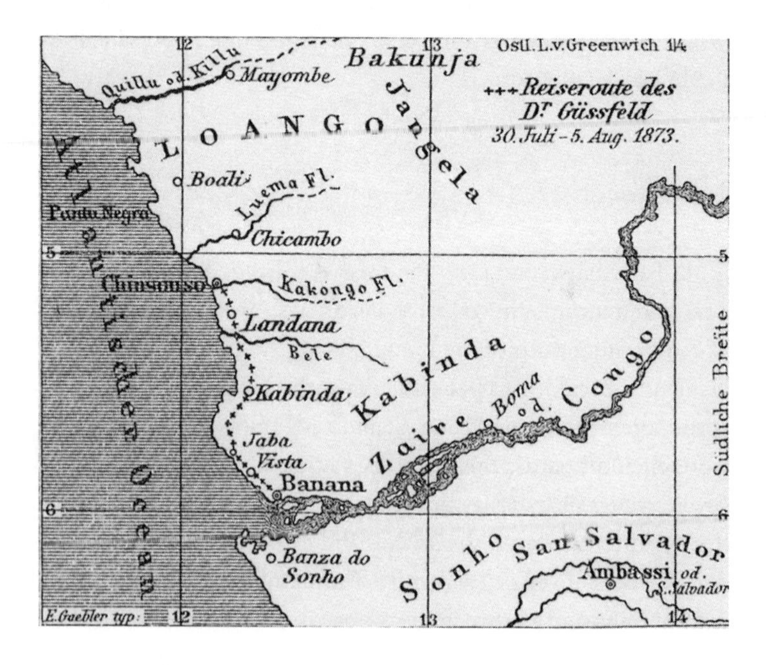

> While the already completed, low dwelling house, built of papyrus stalks, did offer space and a certain comfort, much more than I had expected, the equipment brought along was considerable, and it became apparent very quickly that more space was needed to store it properly. . . . It was therefore a stroke of good fortune that black carpenters were already busy erecting a hut that could be locked and was secure against theft; they could be employed right away for the additional buildings. Needless to say, it seemed that it would take only a few weeks to carry out all the plans. . . . At that time we still set out to work with joyful anticipation, but for months the sites remained selected and measured, and we were still waiting in vain for the completion of the lavatory. What good was all the scolding and encouragement, all the oversight and supervision? The timbers were hard as iron and brittle, the tools unsuitable and in unpracticed hands, many hours of the day were rendered useless by torrential rain, and the Negro always had endless time. He worked at an unspeakably slow and indescribably lackadaisical pace. If you watched the measuring, planing, and smoothing for a while, you had to turn away if you did not want to completely lose patience.

Words that carry racist overtones today were considered completely normal in the nineteenth century. The "superiority" of European culture was a given, not only among German Africa explorers. In fact, Falkenstein showed more understanding for the indigenous population than did many of his colleagues. As a physician he soon realized how dangerous strenuous physical labor could be under the climatic conditions, and that the carpenters they had hired were probably doing the right thing. Still, it was difficult for him and the other members of the expedition to adjust to the circumstances. Ambitious and with no accurate knowledge of the country, the Germans wanted to realize their goals, come

Chinchoxo station, 1875

what may. Tensions and setbacks were a natural by-product of that situation. While those back home were expecting spectacular successes and the exploration of regions still unknown, the expeditionaries in Chinchoxo were constantly wrestling with the same problems. Often the feared malaria alone paralyzed any spirit of enterprise. Then there were the prolific sand fleas, which, alongside the mosquitoes, turned into an almost unbearable scourge. The dampness during the rainy periods was also hard on the participants: "Dampness, stench, mugginess, a truly poisoned air. Everything is covered in mold, the houses are rotting, everywhere a vile, musty smell. An evil force!," Eduard Pechuël-Loesche noted in his journal on 15 April 1875. And that was not all. More obstacles had to be overcome to make the mission—the departure into the hinterland and the beginning of the exploration of the interior—a success. Since expeditions at that time usually carried along a lot of luggage, porters were urgently needed. But those could be procured only with great difficulty, if at all. Many men were brought to Chinchoxo against their will, and they died of illnesses or used

Participants in the expedition. From left: E. Pechüel-Loesche, O. Lindner, P. Güßfeldt, Major von Mechow, J. Falkenstein

every chance to run away. And so the grandiose plans were undone from the start by the simple fact that nobody could be found to carry the loads through the unknown terrain. It is no surprise that conflicts among the members of the expedition intensified under these conditions. Differences in character and temperament, in physical condition and training, now emerged more starkly, and in combination with disagreements about the route, they triggered fierce clashes.

There is something else that is evident to anyone who reads the records of the expeditionaries: although death was far more commonplace and in a certain way "normal" in Germany at that time, in Africa it seems to have been ever-present. The cruel treatment of the natives by some Europeans, and the tensions within the communities residing in Africa, led to cases of shocking violence.

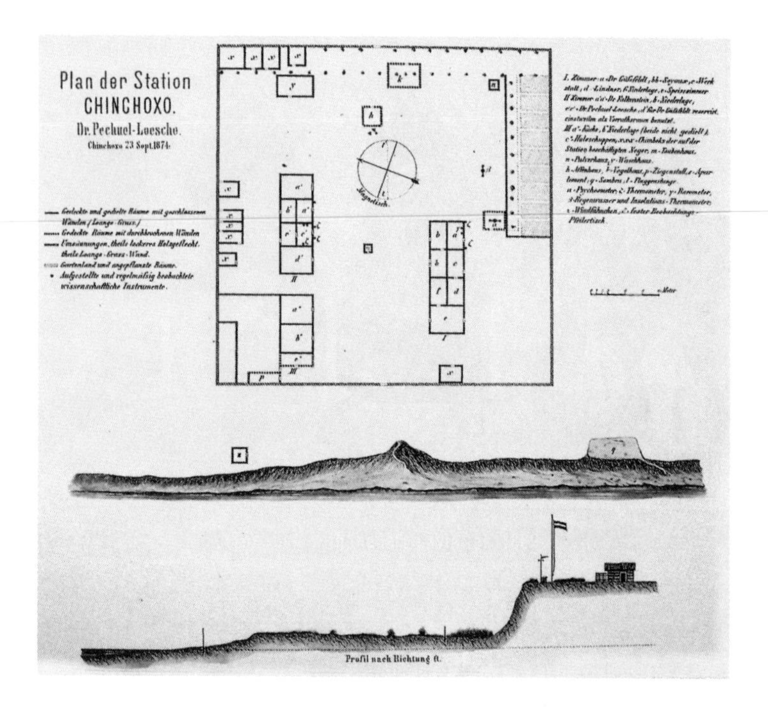

The mentality of an entire region was still imprinted with the slave trade, outlawed only a few short years earlier, and with the tradition of witch burnings. Julius Falkenstein, Eduard Pechuël-Loesche, Paul Güßfeldt, and the botanist Herman Soyaux all had to come to terms with these things. For example, Pechuël-Loesche recorded the events involving the cook of the station, who was accused of witchcraft by his village. Days later, they found his bones and parts of his skull in a fire pit on the beach. But Güßfeldt himself did not shy away from beating the native workers. Embittered as he was about the constant failures and entangled in many prejudices, the mentality of the local population remained alien to him. He considered the need to deal with the headmen of the surrounding villages and to negotiate the terms for the use of the station in prolonged discussions an unreasonable imposition. Finally, when a last attempt was to be made in June 1875 to set out for the interior, nearly all the porters took flight, along with their overseers. Depressed and exhausted, Güßfeldt left Africa only a few days later to deliberate in Berlin about what to do next.

A SMALL "ZOO"

In view of all these challenges, the actual accomplishments of the Loango expedition are all the more remarkable. However, they were different from what had originally been expected. Nearly unnoticed by the German public, numerous chests with preserved animal specimens and dried plants arrived at the museums in the capital beginning in 1874. Masks, everyday objects of the coastal dwellers, photographs, drawings, maps, and measurements demonstrated that the men were not idle. Rather, collecting and documenting was being systematically pursued, and Chinchoxo's meteorological station alone had compiled around forty thousand data items. Since the members of the expedition represented different fields of expertise, it was natural that they all pursued specialized tasks. Such broadly diversified cooperation among scientists had never before existed in sub-Saharan Africa. For example, while Güßfeldt carried out geographical and astronomical measurements and Pechuël-Loesche engaged in linguistic studies, Falkenstein pursued zoological work and was the expedition's photographer. With his heavy equipment, the technically skilled doctor was often at the center of attention, and in some villages he was regarded as a rainmaker. Oftentimes photography, still a young technology at the time, replaced measurements of the heads and bodies of the native inhabitants. Scientists back in Europe were hoping to use such measurements to determine essential human features and differences more precisely, just as much of the data and many of the collections compiled by the expedition were only evaluated back home.

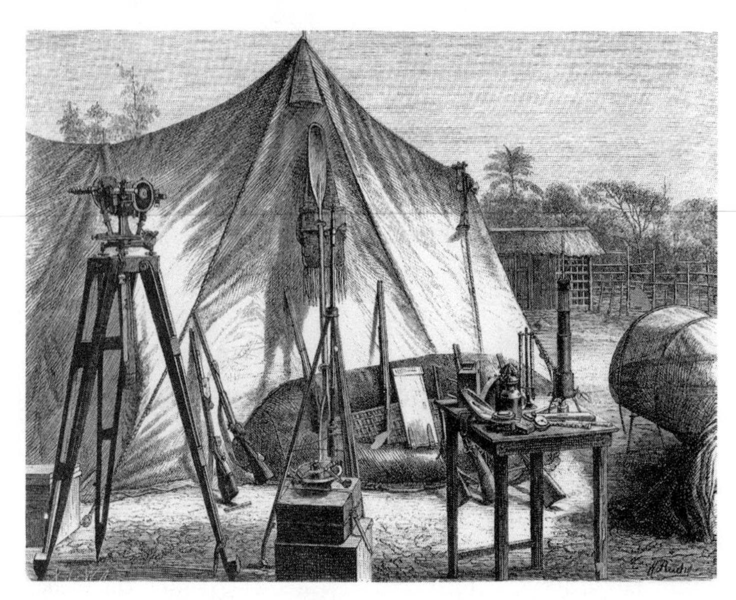

Tent of the German Loango expedition, 1874

As the head of the station and the only doctor far and wide, Falkenstein bore a special responsibility and attended not only to members of the expeditionary group. Other Europeans living in the region, mostly Portuguese, English, and Dutch, were also hoping to obtain his help. "Until our arrival," he wrote in his account, "everybody throughout the entire stretch from Banya to the Congo had been his own doctor, treating himself, following the traditional template, with emetics and laxatives, quinine, opium, and silver nitrate. . . . With the arrival of a German doctor on the coast, this was now instantly different. . . . Everybody wanted to be examined and advised, was satisfied with the medication prescribed once or wrote about a renewal even weeks later, returned home after several hours of rest or, as attested by the extensive baggage, had the firm intention of waiting for recovery on site."

After Falkenstein's arrival in Chinchoxo, the station was gradually expanded, eventually consisting of two residential houses, a kitchen, and several service buildings. The local workers and translators lived in ten huts that were located within the enclosure. Before long the compound also included a small "zoo." Originally intended to allow for a better observation of the "zoological study objects," it also served the members of the expedition as a place of rest and recreation. In his notes, Falkenstein even went so far as to identify this as its most important function, and he advised against similar undertakings elsewhere because of the costs and expenditure of time it involved. Encounters with the animals kept at the station did indeed make it into the later publications by the expeditionaries. In a separate chapter, Herman Soyaux described the aviaries and cages as well as the centrally located monkey house. On a literary walking tour of the compound, the botanist first greeted a gray parrot, who, as expected, squawked "Good morning, Soyaux" in reply. The reader learned that the station was home to about a dozen monkeys, including baboons and long-tailed monkeys, as many as three chimpanzees, several tortoises, and

Animals of the Chinchoxo station, 1875

a tame side-striped jackal. Two rock pythons and a rhinoceros viper were kept in a crate fitted with bars, and a forest antelope was housed in a small mud hut. Birds were especially numerous, since Falkenstein was intensely interested in ornithology. This colorful hodgepodge of wild animals, complemented by sheep, goats, and the two German shepherds, was a constant source of surprises. The old and—by unanimous consent—"exceedingly ugly" female chimpanzee Pauline attracted particular attention. The same was true for the Moor, a universally beloved mangabey (a family related to long-tailed monkeys). She fell overboard during the passage back to Europe, a loss that deeply affected Falkenstein and depressed the mood on board the ship for days.

CHAPTER 4

A Valuable Present

GORILLA HUNTS

The special relationship of the expedition members to the animals at the research station did not keep them from trying their luck at hunting. However, this was difficult in the high grass around Chinchoxo. The only time the Germans managed to see big game animals at all was on forays into the hinterland on the occasional research expeditions. Falkenstein, too, participated in one of these expeditions, which would turn into the most dangerous experience of his years in Africa. For days he was laid up with fever in the jungles along the Kouilou, racked in the end by a bloody cough and mortal terror. Still, the venture undertaken together with Eduard Pechuël-Loesche was a success. Unlike on the open plateau along the ocean, it did not take any great effort to find interesting tracks and clues. The two men shot hippopotamuses, monkeys, and hundreds of birds. Any time a species was unknown or the

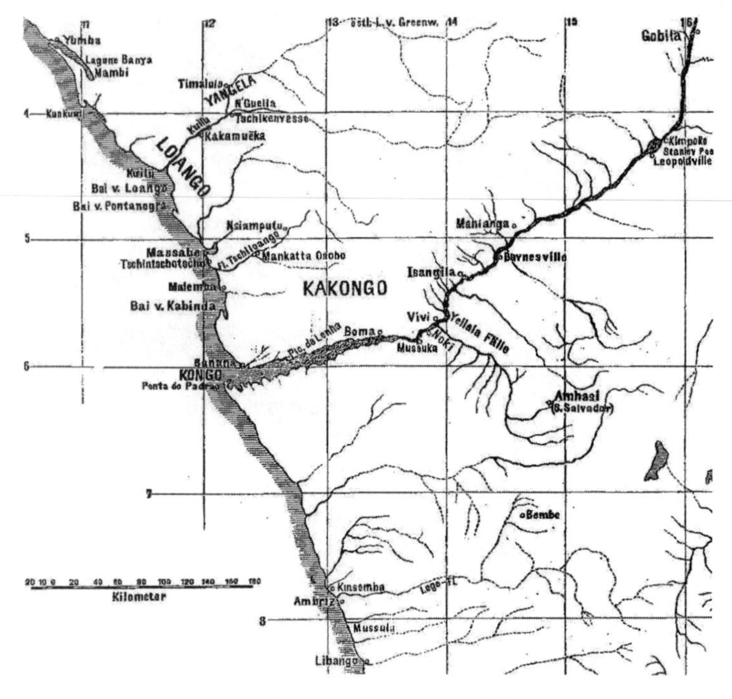

Map of the expedition area, ca. 1880

specimen was especially splendid, it was preserved and carried along in the baggage. In so doing, the men were also fulfilling a request by the board in Berlin, which attached special importance to zoological specimens. Its instructions to also send living "chimpanzees, gorillas, crocodiles, young hippopotamuses, and elephants" to Germany elicited incredulous head-shaking at the research station. Evidently those back home had no idea of the difficulties the expedition was contending with.

While the barking of chimpanzees was one of the recurring experiences of the explorers, the largest of the African apes remained a mystery. *The Homeland of the Gorilla* was the title Eduard Pechuël-Loesche gave to one of his impressive watercolors. Readers who had an opportunity to examine the painting later in several books could make out dense jungle foliage entwined with vines. Even if it appears less menacing in the colored original than in a black-and-white reproduction, it remains a gloomy scene (see illustration on page 45). "Jealous and hostile, the forest fortifies the entrances to and boundaries of its realm," we read in a different passage. Following in the footsteps of Du Chaillu and the German gorilla hunter Hugo von Koppenfels, who had been traveling through Africa since the end of 1873, nearly all members of the expedition tried to track down gorillas. The results were dispiriting. Only the skulls of the animals, venerated as fetishes by the village communities, were collected—if their owners gave permission—and sent to Europe. Interviewing the native hunters yielded very little that was new. "The less the natives know about the gorilla, the more they have to relate about it," one of the explorers remarked sardonically.

Still, another clue can be found in the bulletin of the German Society for the Exploration of Equatorial Africa. Oskar Lenz, traveling in Gabon in 1875 as a geologist on behalf of the Society, reported about a young animal that lived several weeks at a branch of the Hamburg trading house Woermann. Fearful of bites, its keepers had filed down its canines and shackled it to a long chain. Its sleeping area was in a large barrel that was wrapped with a thick sailcloth at night. The approximately two-year-old gorilla, who, as Lenz wrote, became quickly accustomed to the new environment and humans, was supposed to be given as a present to the Zoological Garden in Hamburg. As so often, the transport failed. The gorilla died on the second day of the ocean passage and arrived in the German port city in a barrel of ethyl alcohol.[1]

PONGO

In the end, it was Julius Falkenstein who succeeded in a surprising way. It now redounded to the doctor's benefit that he was

The Loango coast. View from the hills of Chinchoxo to the north, 1875

held in such high esteem among the Europeans of the coast of Loango. That alone can explain why, on 2 October 1875, the Portuguese Laurentio Antonio dos Santos presented him with a young gorilla. This truly astonishing and generous gesture was to thank the German for the selfless help he had offered in several cases of illness. At the time the gift was made, Falkenstein was returning from his trip to Kouilou and was already aware that the results of the German Loango expedition were being judged very harshly back in Berlin. Probably for that reason, as well, he kept silent in his letter about the circumstances under which he had been given the ape at the Ponta Negra trading station. What was needed now was news of successes. The first report about M'Pungu—the name given the gorilla by the natives—is a fascinating document.[2] In contrast to the accounts by Du Chaillu, who had characterized the young animals as wild and vicious, Falkenstein's language is almost tender, and he did not hide his affection:

As the main spoil of the journey that brought Dr. Pechuel Loesche and me into the Quillu region [Kouilou region], one must list a young, living gorilla, who has now been at the Chinchoxo station for more than six weeks under my special care. There is nothing in any way unpleasant about its external appearance, and it would bring much pleasure especially to anyone who loves children. Its physiognomy has something startlingly human-like. The melancholy expression of this black, round child's head, which is always lost in thought, causes the observer to step back with a sense of unease, while the vigorous manner in which he unwillingly rejects every demonstration of affection counsels caution, and the pronounced portliness, the outward-bending knees, and the fairly pronounced flat feet evoke an involuntary smile.

With the exception of those areas that are also completely hairless in humans, its body is covered with a lovely fur, the individual hairs of which are black at the tips, but otherwise dark brown. But at the site where the completely absent tail would be attached there is a white-haired spot the size of a walnut.

All in all one can describe him as indolent, lazy, and disagreeable, with a little imagination as dignified, reasonable, and self-confident. Now and then he gives off plaintively reproachful sounds, which degenerate in the heat of the moment into unpleasant screeching.

Since his primary occupation is quiet and undisturbed sleep, he searches eagerly for an inviting resting place, which he seems to find in the lower portion of my mosquito net. Unfortunately, because of his childish habits and his peculiar, pungent smell, his idea of cleanliness is not in tune with my own, especially since he wraps himself thoughtlessly in the net and thus offers the mosquitoes unfettered access to me. A bed was then set up for him in the form of an elongated, narrow box, which was padded with sacks and furnished

Pongo in Africa

with its own mosquito net. He soon noticed the advantage of the new house, lay down in it in the position in which he was likely born, and escaped the bothersome noise of the outside world with the help of a blanket. After an enviable slumber, the same appetite sometimes tempted him once again to seek out the forbidden spot, either to patiently wait with folded arms for me to wake up, or, rising up in the bed, to gently tug at my sleeves with his small black fingers with their delicate nails to get me to open up the door to his feeding spot.

The food itself is difficult to find, and the Negroes of the house are still trying in vain to procure the various fruits of the forest. His only nourishment now is red grapes that grow on the trunk of a tree and some goat's milk. To enjoy the former as comfortably as possible, he gets a pad from somewhere and then sits, calmly plucking one berry after another and amassing mountains of skin and seeds around him. His favorite pillow is a map of Africa, on which the fingers full of red fruit juice trace unknown routes.

I tried to ease him over the lack of diversion with cheerful music by coaxing the most varied sounds from a glass harmonica. However, he immediately began to blink his eyes, to yawn, and, finally, unmistakably to sleep. After I had produced this effect several times, I desisted and tested whether he would derive any interest from his own self by holding a mirror up to him. However, he took no notice

> whatsoever of his own image, nor of a lively long-tailed monkey I gave him as a playmate.
>
> I am the only person to whom he feels drawn. He seeks refuge with me when something frightens him, which happens easily given his skittish nature. Even the falling of raindrops onto the roof or [an] escaped monkey of the house paying him a visit causes him to scurry to me screeching and on all fours, with his fingers and toes curled in, clasping my legs and reaching up to be taken into my arms. But that does not work anymore when, after going along out of pity, on two occasions he did not want to get down again and tried to hang on with his teeth. Still, we live in complete harmony, and if, gaining reason and strength, he gets accustomed to a mixed diet, there is reason to hope that he can embark on the passage to Europe next summer.[3]

ENDLESS DIFFICULTIES

As already noted, Falkenstein's letter has some omissions. For example, there is no reference to the provenance of Pongo, who had been taken by a native hunter from the hinterland to Ponta Negra and sold to the Portuguese man. As we learn from a later report, in Ponta Negra the gorilla had been tied to a platform scale, where he vegetated in a stupor, half-alive, half-dead. Too weak to reach for the fruit offered to him, the animal was a "wretched sight" and "would surely have died in the next few days." For that reason, too, the new owner was in a hurry to take the first photographs. They were made shortly after the return to the research station and are the earliest pictures of a living gorilla that we have.[4] Moreover, Falkenstein's hope that he could some day take the approximately fifteen-month-old male to Europe gives no hint of the difficulties that he and the members of the expedition would still be facing before then.

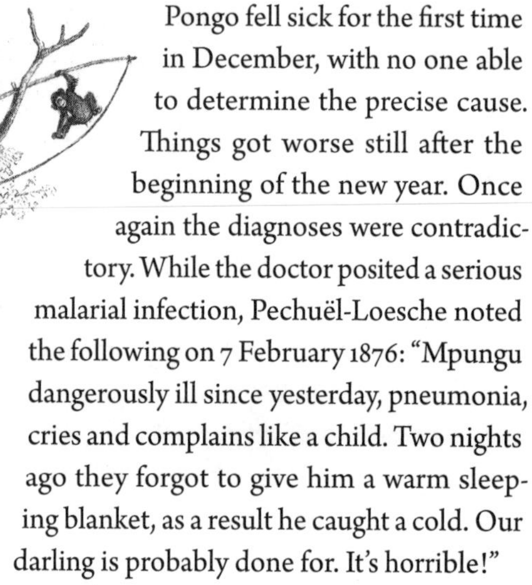

In Chinchoxo, 1876

Pongo fell sick for the first time in December, with no one able to determine the precise cause. Things got worse still after the beginning of the new year. Once again the diagnoses were contradictory. While the doctor posited a serious malarial infection, Pechuël-Loesche noted the following on 7 February 1876: “Mpungu dangerously ill since yesterday, pneumonia, cries and complains like a child. Two nights ago they forgot to give him a warm sleeping blanket, as a result he caught a cold. Our darling is probably done for. It’s horrible!”

The illness, which dragged on for five or six weeks and was accompanied by violent cramps, kept the entire station on edge. Like any child, the gorilla was treated with quinine and calomel, bathed every day, and rubbed with ointments. When his whimpering would not cease, Falkenstein took him into his bed for ten nights. Not until March did the ape’s health improve, and the “quite stiff body” slowly regained its former agility. “Only now do we feel how much affection we have for the animal; we nurse him with the greatest care, if only we can keep him alive,” we read in Pechuël-Loesche, who was, after Falkenstein, the most important figure for the gorilla.

Unlike all the other animals, Pongo spent his nights not in the courtyard or in the monkey house but in Falkenstein’s room, where he slept in the abovementioned box. In other respects he was treated entirely like a member of the household: he sat at the table with the men and thus became accustomed to human food. He also ate bananas, oranges, mangoes, and most of all the egg-shaped berries of the *Annona senegalensis,* a fruit that tastes similar to pineapple. Pongo’s admirers also included the inhabitants of the surrounding villages, who sometimes stopped by to visit the gorilla. One girl from the immediate vicinity of the station came almost every day. Eduard Pechuël-Loesche, who among all the expeditionaries cultivated the best contacts with the Africans, remembered this special bond even decades later. "Our gorilla," the geographer wrote in his Loango book, not published until 1907,

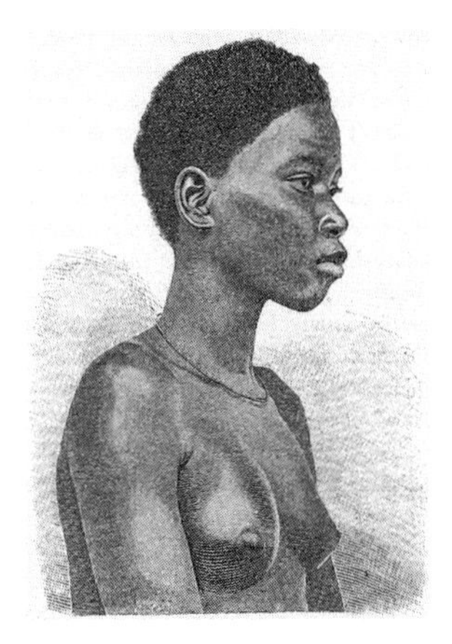

Girl, Loango coast

> had taken into his heart a kind, incomparably charming girl. Her name was Nkambisi, and she possessed to a high degree what we call sunny cheerfulness and noble serenity. The gorilla, as one sometimes observes in animals, was positively enchanted by her movements and especially her voice. And indeed, Nkambisi had an unusually melodious voice, and so supple and expressive that her speech sounded like music. When the gorilla had fallen very ill, she visited him and devoted as much time to him as to a suffering child, nursed him, and delivered long, funny speeches, telling him that he should stop playing around in the courtyard, stop inspecting the market baskets of the women peddlers, and not cause so many worries to his fathers, meaning us.

Since most of the members of the expedition reported on Pongo, we know other details of his stay from various other sources: his surreptitious raids, his penchant for sugar and sweet tea, his games with the long-tailed monkey Moor and his interactions with the female chimpanzee Pauline. The ape was allowed to move freely about the station compound, where ever new "adventures" ensued.

Baobab tree near the research station, 1875

Once he simply disappeared into an adjoining field, with only the moving stalks of grain betraying his presence; another time he fell from a tree, fortunately not a very tall one; or he used an overturned bucket as a drum to summon the neighborhood. It is these everyday stories, which the authors are so fond of recounting, that make one thing clear: Pongo had become used to the conditions in Chinchoxo. Torn from his natural habitat, the gorilla had "imprinted on humans." Good preconditions, then, for hazarding the passage to Europe, which had been planned for some time and yet evoked some bitter feelings. After all, the men could not expect a triumphal reception in Germany. The expedition, which had been launched with such high expectations, was now officially considered a failure, undone by the unfavorable circumstances and the insurmountable difficulties involving the issue of the porters, and undone by the impatience back home and the financing question. Pongo was therefore all the more important, the "most valuable prize" of this pioneering venture of German exploration in Africa. Having set out to survey a continent, at the end of their journey the men were left with a small gorilla to offer proof of all their exertions and to attract attention one more time.

CHAPTER 5

Darwin's Felicitations

ON THE POSTAL BOAT

News that the expedition was being terminated had cast a pall over the mood of the participants. For all the hardships they had suffered, the explorers felt they were being torn away from their work, and they had grown fond of the station in Africa. The farewell was accordingly melancholy and dragged on for days. Everything had to be packed up or sold on the spot. And yet a lot of new baggage had accumulated. In addition to the scientific instruments, the specimens, and the personal effects, a small menagerie was setting out on the journey home. The monkeys and the tortoises, the forest antelope and the chimpanzees were put into barred shipping crates, while Pongo remained with Julius Falkenstein. The agent of the Dutch trading station hosted a farewell meal in Landana, the closest town. The cargo was loaded into boats on 5 May 1876 and taken to the postal ship waiting off the coast. This transfer alone was not without risks on account of the surf. "We ourselves climbed into the last boat," wrote Herman Soyaux. "On the beach stood our white friends and the Negroes who had come from Lusala and other neighboring villages and waved to us. Clothes were waved, flags fluttered, gun salutes roared. . . . It was eleven in the morning when we came aboard, and shortly after the anchor jerkily clanked up out of the depths."

The boat that was to take the four men and their baggage to Europe was called the *Loanda*, and, as on the initial journey, it belonged to the British and African Steam Ship Company. Built around 1870, the *Loanda* was a steamer eighty-five meters long and ten meters wide, with a speed of around ten knots. Although the *Loanda* also carried mail, it served chiefly to transport palm oil and other goods. As there were only a few cabins for passengers, traders taking the boat for shorter trips often slept outdoors. But Falkenstein, Pechuël-Loesche, Soyaux, and Lindner were settling in for a passage that would take weeks: the baggage was stored away, personal belongings were unpacked, and the animals looked after in steerage. Being under Falkenstein's care, Pongo was counted among the cabin passengers. But the living charges of the Germans were not the only animals on board. Far from it. To begin with, the new passengers aboard the ship had to get used to the noise of hundreds of gray parrots and other birds on the quarterdeck. These belonged to the sailors, who had acquired them in African ports. Like numerous monkeys and guenons, pangolins, porcupines, mongooses, and giant squirrels, they would be sold upon arrival and earn the sailors some extra cash. Even though this trade was prohibited by the shipping company, the captain did not begrudge his crew this business, as long as it did not interfere with their day-to-day duties.

The captain of the *Loanda*, James Clancy, proved a godsend for the returning explorers. Affable and friendly, he acceded to their wishes and treated Pongo with great consideration. On his orders, all objects on board that could injure the gorilla were checked and tied down, and the stewards and kitchen staff looked after the needs of the precious passenger. The little ape was allowed to join dinner in the ship's salon and move freely on the deck. There he was given a "strong rope for gymnastic exercises" and a bunting "for protection against the northerly breeze." To keep him from looking over the side of the boat with excessive curiosity, Falkenstein held him over the railing for a few minutes, a lesson that worked wonders. Since the expedition's monkeys were also taken out of their boxes on occasion and tied to the ship's shrouds, at times the steamer resembled a floating primate station.

Like all postal ships of the two English lines that traveled from Liverpool to western and central Africa, the *Loanda* called on about a dozen ports, and if the surf was heavy it would wait for days to load cargo. While that slowed down the trip, it also made it more interesting. In his *Aus West-Afrika, 1873–1876: Erlebnisse und Beobachtungen* (From West Africa, 1873–1876: Experiences and observations), Herman Soyaux described the journey along Africa's west coast, painting a richly colored picture over many pages. Time and again it is Pongo who attracts the greatest attention. In Calabar (in modern-day Nigeria), the crew even disembarked with the gorilla to attend a celebration on one of the hulks, a houseboat used by the Europeans. If the reports are to be believed, Falkenstein's charge felt extraordinarily well during the outing, giving the best performances unbidden when most of the guests were assembled.

Once the travelers had passed the Canary Islands, the weather slowly cooled. In contrast to the monkeys, this did not seem to bother Pongo very much. His state of health unchanged, he moved across the deck, entertaining the passengers who had come aboard in Gran Canaria, including five children, with unexpected antics. "Everything as it was! Gorilla is tolerating the climatic changes very well, is quite wild with high spirits," Eduard Pechuël-Loesche recorded in his journal on 9 June. Eleven days later he ended his log with the key words, "This morning coast in sight; very cold; sharp wind from the east. Many seagulls!"

BACK IN EUROPE

On 21 June 1876, the *Loanda* steamed into the enormous port facilities at Liverpool. This marked the end to a strenuous journey for the men of the expedition. While still on board, Falkenstein and Pechuël-Loesche had engaged in long conversations about Africa, and the talk kept coming back to Chinchoxo. But now there was little time to indulge in reminiscences. Barely on land, they had to supervise the transfer of the baggage and the care of the animals. The small menagerie had suffered no losses apart from the mangabey, which had jumped overboard, and the forest antelope, which had died during the very first days. The gray parrots and the other sailors' animals were now also taken out of their cramped crates. The seamen did not have to search long for buyers. Many people living in increasingly crowded cities were yearning for nature. Exotic birds and small monkeys were in high demand, even if the vast majority of animal housemates was made up of dogs, cats, or ornamental fish. Never before had so many animals inhabited the parlors of the bourgeoisie or had so many organizations been devoted to their breeding and keeping.

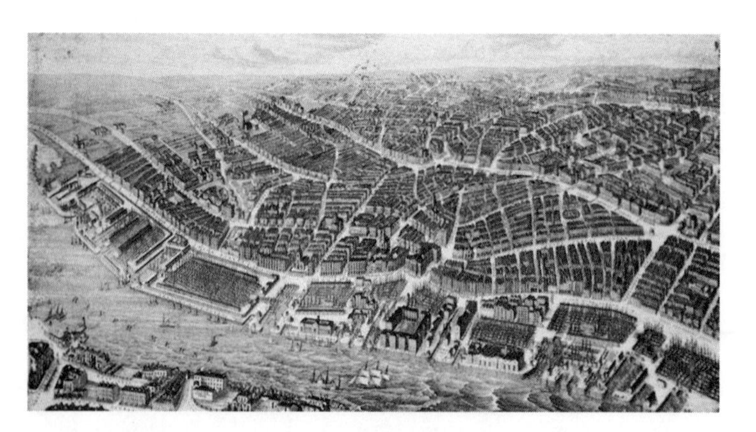

Panorama of Liverpool, ca. 1850

Like many Germans who set out for the world from Liverpool or arrived at its port, the returning explorers took accommodations in Eberle's Alexandra Hotel. This place, close to the train station, was considered reasonably priced, and the German-speaking staff made it easier to get one's bearings. The next morning, the *Liverpool Mercury* published the *Loanda*'s passenger manifest. It would have been easy for readers to overlook the fact that a small gorilla was mentioned at the end of the short list. But by this time, word of Pongo's arrival had already gotten around. And with every hour that the ape spent at the English port, interest in him seemed to be growing. This is how Falkenstein remembered the circumstances weeks later: "The entire prodigious city was buzzing with excitement. Dense crowds surged back and forth at the windows of our hotel, while the better off, among them the leaders of the city and the top scholars, called on us in person, and the aged Darwin conveyed his congratulations in writing."[1] We do not know whether the otherwise rather sober and pragmatic doctor was overcome by his own emotions on this occasion. In any case, a series of other articles appeared after the arrival of the ship, with the *Hampshire Telegraph*, the *Liverpool Post*, and other papers publishing shorter reports. Thomas J. Moore, the curator of the Liverpool Free Public Museum, wrote a piece in the *London Times*, one of the country's leading papers. At his request, Pongo had been brought to the museum and confronted with a stuffed gorilla. To the disappointment of the scientists, the encounter yielded nothing. Pongo could not be prompted into any show of emotion. Instead, he clambered about between Falkenstein's legs and had his fun with those present.

"The animal," the correspondent of the *Liverpool Mercury* wrote on 24 June,

> has a large head, a massive chest, and very strongly formed limbs. It is a most playful creature, and greatly resembles a child in the way in which it fondles about those in whose

Free Public Museum in Liverpool, ca. 1860

care it is. It appears to have become strongly attached to Dr. Falkenstein. It climbs round his knees until it reaches his arms and then when lifted up embraces him round the neck and waist just as a child about the same age would do. . . . Sometimes it throws itself round a person's legs and apparently attempts to use its teeth, which are formidable-looking and well set; but after giving a slight pinch it releases its hold and it appears to be only the creature's fun. It dislikes to be alone, and when it was placed in a room by itself it made a great noise until released. Its principal food is rice and milk, but it does not object to a good slice of roast beef. It can walk upright, but when it wishes to get along faster it goes on both hands and feet, by a sort of sideways movement, much in the same way as one of the human species would do if going on all fours. Its eyes, which are dark and sparkling, have a cunning twinkle about them when the animal is at play, showing that it carefully watches every movement, from the way in which it suddenly shifts is position. Its nose is broad, the nostrils being very wide; and its mouth is large. When pleased it gives a peculiar grunt.

Other observers were also impressed and predicted that Falkenstein's gorilla would attract "worldwide attention." The ape was not only of "great scientific interest," wrote the reporter from the *London Times,* but also demonstrated a highly engaging nature. And while the British papers were still lamenting the imminent departure of the "rare anthropoid," the members of the expedition were already setting out for home. After four days they left Liverpool and Pongo was given a free train ride to the port city of Hull, where the steamship to Germany was already tied up on the quay.

INTENSE CONFLICTS

Two days later, an illustrious crowd was gathered in the Hanseatic city of Hamburg. Geographers and zoologists, journalists and the curious—all were awaiting the returning Africa explorers. Otto Hermes, the director of the Aquarium "Unter den Linden," and Heinrich Bodinus, his colleague from the Zoological Garden, had come all the way from Berlin. The Leipzig animal painter and writer Heinrich Leutemann, too, did not want to miss the gorilla's arrival. On 27 June he visited Zingg's Hotel, where the travelers had taken lodging the previous evening. When he arrived, Pongo was gamboling on top of the unlaid tables in the dining hall, and under the gaze of hundreds of visitors he had to be repeatedly "called to order." But the real struggles of those days took place behind the scenes, for the question of where the ape would be exhibited in the future was far from settled. All that was clear is that, for "reasons of national interest," he was to be shown in Berlin. Heinrich Leutemann reported on a veritable "gorilla battle" on 12 July 1876 in the *Leipziger Tageblatt.* "Here Aquarium, there Zoological Garden, thus resound the battle cries," we read in an article about the events of those days.

The issue also involved legal questions. Legally, Pongo belonged to the German Society for the Exploration of Equatorial Africa. Like

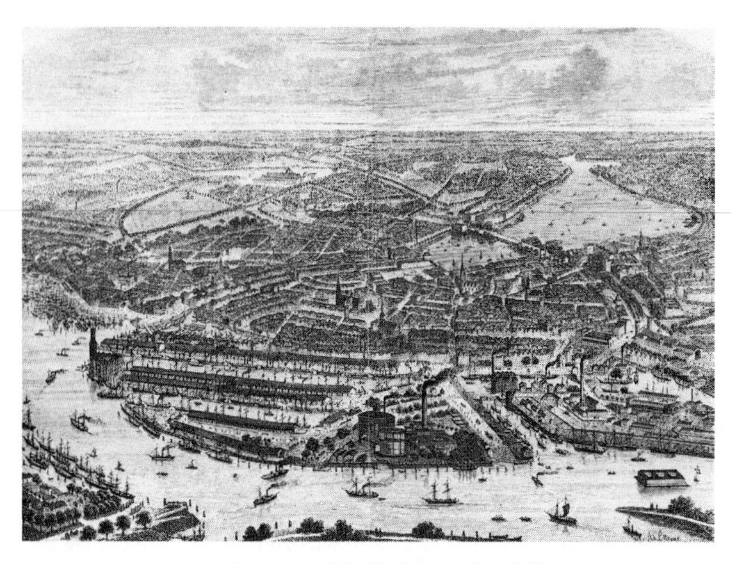

Panorama of the Hamburg harbor, 1882

nearly everything the travelers had acquired along the way, he had passed automatically into their possession. And so while Falkenstein could participate in the discussion, he had to leave the final word to others. Large sums of money were at stake in the decision about the fate of the animal. In Liverpool a businessman had offered six hundred pounds for the gorilla, and in Hamburg, too, he was coveted. While the gorilla was still at Zingg's Hotel, the animal merchant Carl Hagenbeck made the astonishing offer of twenty thousand marks. Never before had anything like this been paid for an ape. For example, young chimpanzees rarely cost more than twelve hundred marks,[2] more or less equivalent to the annual income of a mason at the time. Although Otto Hermes matched Hagenbeck and raised his offer, no agreement could be reached in Hamburg. Pongo would be turned over to the Berlin Aquarium only for the time being and until the final decision was made by a guardianship committee.

Given these circumstances, the intensive talks, and the secretive machinations, the stay in the Hanseatic city became almost secondary, even though a worthy reception had been arranged for the members of the expedition. When Julius Falkenstein and Eduard Pechuël-Loesche presented themselves to the city's Scientific Society, those present rose from their seats in honor of the returning explorers. And once again all eyes were on the gorilla. It remained that way when Pechuël-Loesche spoke about their sojourn on the west coast of Africa and described "in vivid images the life story of a Loango Negro from birth to death." After the lecture, a review of which appeared the very next day in the *Hamburgischer Correspondent*, the travelers began to pack a second time and to say their farewells. While Otto Lindner remained in Hamburg, Julius Falkenstein boarded the train to Berlin, where he arrived on 30 June 1876, around 6:45 in the evening.

Gustav Nachtigal

The large hall of the Lehrter train station now provided the stage for an unexpected scene: instead of handing Pongo over to Gustav Nachtigal, a famous Africa explorer and the new chairman of the Society, Falkenstein refused to part with his beloved animal. Surprised and irritated, Nachtigal tried to keep his composure. Instead of the touching pictures the journalists had been expecting, they were treated to a loud and heated argument. The gorilla, Falkenstein asserted, was exhausted from the excitement of the last few days, which is why a sudden change could harm him. The handover would therefore have to be postponed. In the end, Otto Hermes had no choice but to let the new star of the Aquarium go. We learn from reports in several papers that the doctor spent the night in the Hotel de Magdebourg, where Pongo was the source of excitement into the evening hours. Since Falkenstein's journals have been lost, we do not know exactly why he acted this way. Was he giving voice to his disappointment over the order to terminate the expedition, or was he concerned solely about the well-being of his African travel companion? The ape was clearly used to moving in the open air. Any change for the worse in those circumstances could have had dangerous consequences. It was only the next afternoon that Falkenstein climbed into a carriage and brought the gorilla to the Aquarium "Unter den Linden."

When it comes to the subsequent negotiations, as well, all we have are speculations. It would appear, in any case, that the talks preoccupied Gustav Nachtigal so much that he barely found time for anything else. We read this in a letter dated 9 July:

> If you, like me, had lived through two gorilla weeks, with all their excitement and vexations, their hopes and fears, you would extend forgiveness to anyone guilty of having neglected his obligations toward you. . . . A fortnight ago

Lehrter train station in Berlin, 1879

> today, the noble creature, having completed its sea voyage from Liverpool, set out for his new home. Today it is ten days since it arrived at the Lehrter train station, and only yesterday I signed the agreement that transferred our cousin definitively to the Aquarium, on the condition of constructing new, airy premises for him, assigning him a permanent companion, and always following the directives of the guardianship committee with respect to his care and upbringing. But the struggles and difficulties we had to go through to arrive at this end! First they wanted to sell this first specimen of its species secretly in England for £600, then cunningly in Hamburg for 20,000 marks, then Dr. Falkenstein refused to hand him over, since he believed that we intended to kill him by transferring him to the Aquarium.[3]

This letter, addressed to an unknown lady, proves two things. First, the African Society had ensured that Pongo had a guardianship committee, whose directives nobody could easily ignore. With Rudolf Virchow, Christian Gerlach, and Robert Hartmann, that committee was staffed with prominent members, and it also included Julius Falkenstein and another doctor. Second, the agreement entered into with the Aquarium contained concrete stipulations that had to be implemented now. Falkenstein initially helped in familiarizing the ape with his new surroundings. He did not have much time for this task, as he was soon posted to the Prussian fortress of Gaudenz. Although the military doctor visited Pongo on occasion, others assumed the responsibility now. Henceforth, the gorilla had to get used to Otto Hermes and the conditions at the Aquarium. The fact that he was also given two caretakers who looked after him day and night was part of the agreement and may have made it easier for Falkenstein to take his leave.

CHAPTER 6

The Aquarium "Unter den Linden"

LEARNING BY SEEING

The building in which Pongo was now housed did not look very spectacular on the outside. If one did not know what was located behind the façade, it would have been easy to overlook, and this even though the Aquarium "Unter den Linden" was the most unusual display that Berlin had to offer at that time: a world of fantasy, yet molded out of rock. On just two levels, the visitor was greeted by a world full of grottoes and boulders, coves and spiderwebs woven from iron. Here a serpentine path beckoned, there a waterfall plunged into the depths, and at the very center lay a fairylike aviary, filled with the chirping of hundreds of birds. In semidarkness the curious completed the three-hundred-meter-long path through the complex, past the large water basin and around one hundred cases and cages. No wonder that some visitors were reminded of a fairy tale, while others thought of a mine or the entrance to the netherworld. And yet all this extravagance was more than merely an effort to impress the observer, for behind this stony magnificence was a very specific intent. The creators of the Aquarium, the zoologist Alfred Brehm and the architect Wilhelm Lüer, wanted to educate and convey knowledge through observation. That they employed such daring scenery for that purpose was not only a response to the desires of the public. As in his *Illustrirtes Thierleben* (*Illustrated Life of Animals*), which swelled in its second edition to ten volumes weighing in at seventeen kilograms, Brehm went all out at the Aquarium. The variety of life was to be revealed in this building and displayed in what was really not a great deal of space. The underlying idea—that animals and plants, rocks and geological formations were given space side by side—was correspondingly ambitious. Visitors walked through various climate and vegetation zones, finding themselves sometimes above and sometimes below the sea, and at the very end passed by a reproduction of the Blue Grotto of Capri at a scale of 1:6.

But a walking tour alone is not a sufficient explanation. To fully understand the building, its history is equally important. Brehm had moved to Berlin in 1867 filled with grand visions. It was not an aquarium that he wanted to build here, in the capital of the North German Confederation, but a "zoological garden under a single roof." With a stroke of the pen, the existing plans were discarded and new ideas drafted. Now the facility was also to house lizards and snakes, birds and smaller mammals. At the same time, the ambition to display sea animals in a country's interior was spectacular enough. A seawater aquarium demanded expertise and major investments. Only fourteen years ago, the "Fish House" had opened its doors in London, delighting and astonishing the visitors with its colorful denizens. Although the small "water menagerie" was still very modest and more like a greenhouse from the outside, it caught

City plan with the Berlin Aquarium, ca. 1878

Alfred Brehm

on quickly. Similar and far more spectacular institutions were gradually constructed in Paris, Vienna, Brussels, and Rotterdam. Brehm, who as the former director of the Hamburg Zoo had practical experience in the field of aquarium science, was inspired by them. Supported by the founding committee and the stock corporation established soon after, he planned to build "the largest and most important aquarium of the world" in Berlin. Land was purchased, and—equally important—a different architect was hired: Wilhelm Lüer. For unlike the doyen of aquarium construction, the Englishman William Lloyd, Brehm was all about the staging of habitats. A similar idea animated the work of Lüer from Hannover, who had erected several enclosures and buildings for the city's Zoological Garden. The grotto aquarium he designed for a businessman so fascinated Brehm that he invited Lüer to join the project.

MASTERFUL TECHNICAL FEATS

The plot of land acquired by the stock corporation was at the corner of Unter den Linden and Schadowstraße. Its central location had been the decisive factor in its purchase. There was hardly a Berliner who did not pass by here at some point, and no tourist would miss seeing Prussia's grand avenue. However, the building site itself, in the bed of a former branch of the Spree River, had to be prepared at considerable expense. Situated in the rear of several buildings, it was surrounded by the firewalls of the surrounding houses. Henceforth, its Z-shaped footprint determined the architect's plans. A labyrinth of passageways and grottoes was built on several levels, with more than one hundred wagon loads of natural stone (each weighing ten thousand kilograms) used in its construction. Granite and basalt, quartz and Jura limestone, greywacke and slate, marble and erratic blocks . . . the loads were rolling in from all over Germany. Equally impressive were the technical details of

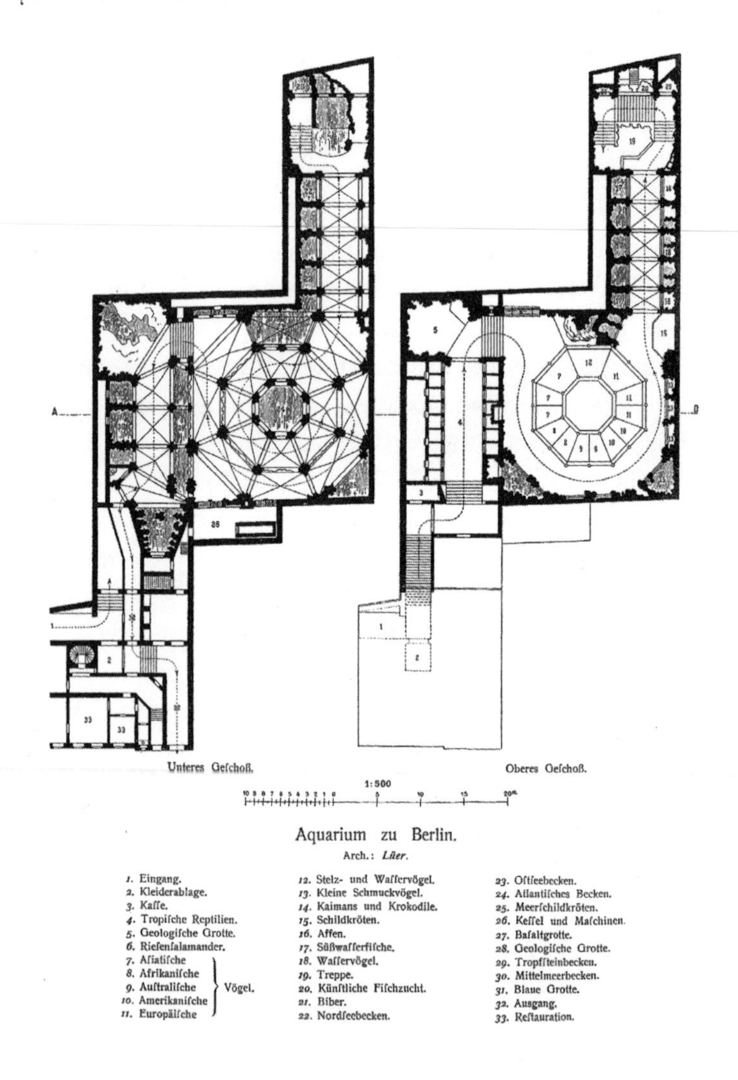

the building. Since no electrical grid existed as of yet, the upper level, where daylight entered through the glass roof, was intended for the terrarium animals and the birds. Light shafts and a strong glass plate that closed off the middle room of the large aviaries allowed for natural lighting also in the spaces below. There were no windows in the area of the actual aquarium with its saltwater

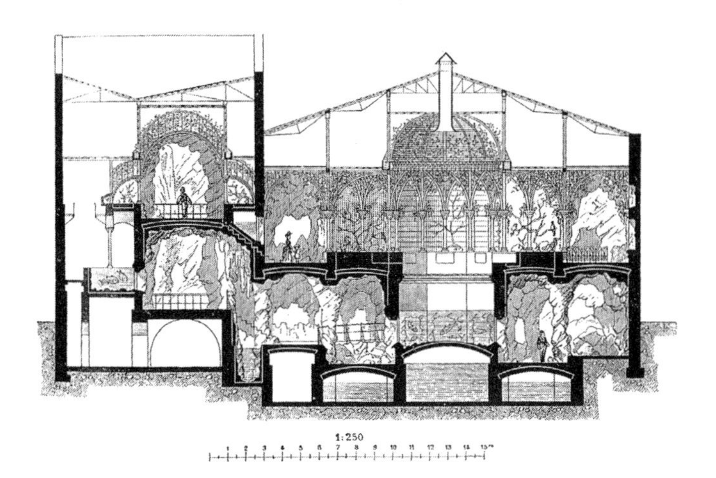

Cross-section of the Berlin Aquarium, 1869

tanks, which is why gas lamps flickered in the visitor walkways. The building was heated by a system of hot water pipes, which also warmed the sand floor of the terrariums. Air circulation was achieved with a ventilator and with open hatches in the attic floor. Located in the basement, which was closed to visitors, was the reservoir of seawater. To suppress the growth of algae, the water was first stored in complete darkness for twenty-four hours without any exposure to air. Then an 8 hp steam machine pumped it into three elevated basins, from where it flowed into the display tanks through an elaborate system of pipes.

Alfred Brehm reported that more than three kilometers of pipes were laid in the building. The Borsig steam machine, the numerous iron structures, the heating and ventilation systems—all of this was state of the art or better. At the same time, the Gothicizing architectural elements for which Wilhelm Lüer had an affinity lent the Aquarium a strangely otherworldly quality. For example, the aviary at the center of the gallery, strung with wire and piano strings, had iron columns that were connected with

The room with the large aviaries, 1869

The residential and administrative building of the Berlin Aquarium, ca. 1875

pointed arches. The visitor was supposed to experience the space as an "architectural jungle" that would induce astonishment and contemplation.

Interest in these kinds of attractions was undiminished. In dozens of journals and even more organizations, laypeople and university professors were discussing natural phenomena and the latest scientific insights. Excursions and lectures, exhibits and conferences on biological or zoological topics were the order of the day. There was good reason for many writers to believe that the "general participation of humanity in the discoveries of science" was an important characteristic of the time. Those living in that century looked forward to the future with optimism and firm confidence. Brehm, himself one of the most important popularizers of scientific knowledge, recognized this early on and tailored his projects accordingly.

TWO DIRECTORS

The stock corporation also owned a corner house fronting the street, which became a hotel in later years. It was here, in the offices of the administration, that many of the animals intended for the Aquarium were initially housed. In August 1868, eleven months after the cornerstone of the Aquarium had been laid, the topping-out ceremony was celebrated (a ceremony to mark the placement of a building's highest beam). The more news reached the outside world, the more curious the public became. In contrast to the Berlin Zoo, which had been considered a problem child for a number of years, the new institution seemed to confirm the wildest hopes. "This zoological garden under a single roof is the most splendid and the largest of its kind in the world. . . . If Berlin wants to call itself a world city, the Aquarium gives it the right to do so," wrote the *Illustrirte Zeitung* even before the official opening.

Otto Hermes

When Prussia's King Wilhelm I officially dedicated the building on 11 May 1869, the event attracted attention far and wide. Along with the king's favor, the crush of visitors spoke volumes. In three months, the company counted more than one hundred thousand attendees, a success that filled all those involved with pride. Nevertheless, Brehm spoke openly about a number of things that still worried him. Some windowpanes had cracked during the first few weeks, and there were still gaps in the animal stock. Above all, the seawater was a source of trouble. Initially delivered in barrels, it turned "milky," causing visitors to peer into a white liquid. It was months before the chemist Otto Hermes, who was consulted about the problem, was able to come up with a remedy. His procedure for producing artificial seawater was an innovation that was taken note of around the world. The signs announcing that the water was "in the process of being clarified" disappeared, and Hermes, who had initially worked on a fee basis, remained closely affiliated with the institution. In the summer of 1871, he became deputy director and, like Alfred Brehm, a personally liable partner in the stock corporation. To what extent the founder of the Aquarium felt slighted by this move is a question that exercised the newspapers at the time. In any case, tensions existed. Brehm was not the kind of man who would put up with anyone telling him how to do his job. Temperamental and opinionated, he was considered a complicated man by his contemporaries. As much as he loved audacious projects, he had little patience for the day-to-day operations. Instead, he worked on a new book, went on frequent trips, and developed further and even more grandiose plans. It is perfectly understandable that the stockholders were concerned and saw in Otto Hermes someone they could trust. But the collaboration between the two men proved difficult and eventually ground completely to a halt. At the end of a long period of turbulence and mutual recriminations, Brehm was forced to step down, given leave by the supervisory board, and terminated effective 1 April 1874.

In Otto Hermes, the director's job was given to a man who quickly achieved recognition. As he oversaw the operations of the corporation for more than thirty-six years, the history of the Aquarium is closely tied to his biography. Born the son of a landowner in the Prignitz region, Hermes studied pharmacology and natural sciences and established his own businesses in 1864. His chemical factory in Berlin was located in Ritterstraße, in the modern-day district of Kreuzberg. With a doctorate in chemistry, Hermes was a man of many talents. He published work on sulfur and sodium compounds, took an interest in zoology and fish breeding, and entered politics early on. As a liberal, he fought on many fronts and was a member of the Prussian Landtag and the Reichstag.[1] But in the summer of 1876 all that was yet to come, and at the age of thirty-eight his primary concern was running the stock corporation.

The Berlin Aquarium had to compete with many institutions. The fact that the Zoological Garden had carried out a large program of new construction following the creation of the German Reich troubled Hermes. At the same time, though, the institution he led was by now mentioned in every guidebook. With its idea of presenting animals according to habitats, the Aquarium was decades ahead of its time. Since the parcel "Unter den Linden" was steadily increasing in value and the residential house could be rented out, the stock corporation had financial possibilities of which many zoo directors could only dream. Successes in animal keeping and in the breeding of birds and fish, and an excellent

Orangutans, March 1876

assortment of lizards and snakes, attracted attention and were talked about among the experts.[2] But there was another attraction that made the Aquarium truly popular: the exhibition of apes. The fact that no fewer than three species could be seen after Pongo's arrival was considered a sensation: "One gorilla, two orangutans, and two chimpanzees—that is more than even the most insatiable zoologist can digest all at once," the editors of *Die Natur* told their readers in the summer of 1876.

MOLLY THE FEMALE CHIMPANZEE AND AN ADULT ORANGUTAN

This presentation, unique in the world, was by no means planned. Under the conditions at that time, nobody would have dared to hope that they might acquire such a collection of animals. From

In the aquarium, March 1876

the outset, though, the space with the large aviaries had a monkey cage, in which initially other animals were kept. The female chimpanzee Molly arrived in April 1870. Too weak to sit up, she resembled "a corpse on leave" and could not be shown to the public for weeks. That the first ape the Berliners would see in their city did recover in the end was thanks to Carl Seidel, the Aquarium's feed master.[3] He remained by her side day and night and eventually nursed her back to good health. Only eighteen months later, however, a "lung ailment" carried off the young animal. A second chimpanzee was acquired after Molly's death, launching a tradition that Otto Hermes continued for around twenty years. Although the purchase of an ape was a financial risk, it promised to generate a lot of revenue. Moreover, it attracted the public's curiosity and consolidated the institution's reputation among experts. And there was something else altogether: for anyone educated in the natural sciences, taking an interest in apes was a given at the time. Hermes, too, traveled to the Dresden Zoo for the express purpose of seeing the female chimpanzee Mafoka. On 1 March 1876, the director of the Aquarium achieved a major coup. After a nighttime journey to the Alster River, he bought two orangutans from Carl Hagenbeck. Since one of the animals was an adult specimen, the first ever shown in Europe, the animal trader asked for nine thousand marks. Hermes, who had to have the sum approved by the supervisory board after the

Alfred Brehm with the female chimpanzee Molly (name used multiple times), 1866

fact, did not hesitate for a moment. Back in Berlin, the boxes were unloaded by "ten strong men" in the presence of several journalists. The very next day the dailies were publishing detailed reports, and visitors came in droves.

The attention that apes attracted in those years is also reflected in the specialized journals. For example, in the *Zeitschrift für Ethnologie* (Journal of ethnology) the philologist Carl Nissle published an article about Molly's life that ran to eleven pages and included several drawings. Authors such as the director of the Frankfurt Zoo and veterinarian Max Schmidt described their observations of chimpanzees and orangutans down to the most minute detail in long, multipart pieces. A look at these reports makes one thing clear: whether in Frankfurt or Dresden, Berlin or London, the conditions under which the animals were kept were similar. Pongo was thus not a special case. Even if he was cared for at great, nearly unimaginable effort and expense, the personnel were following a tried-and-true practice. Beginning in July 1876, visitors could see that for themselves, as henceforth interest was focused on the gorilla. He, above all others, mesmerized curious visitors and ensured the best financial performance of the Aquarium since its opening.

CHAPTER 7

The Most Popular Resident of Berlin

A SENSATION

"The Gorilla Is Finally Here," "About Our Gorilla," "The Gorilla in the Aquarium"—thus ran the headlines following Pongo's arrival. The *Berliner Fremdenblatt* and other papers even put the story on the front page. The *Tribüne* reprinted two articles from the *London Times,* while the *Vossische Zeitung* sent its own reporter to cover the welcome at the Lehrter train station. Otto Hermes, who had good relations with the journalists, had every reason to be pleased. That the gorilla was described as the "rarest and most interesting, the mightiest and most expensive ape" in the history of zoology was very much in his interest. For in the first few days, what counted most was the sensation. The price paid for Pongo seemed so astonishing that one editor calculated the cost for each kilogram of the gorilla. And when Carl Hagenbeck offered a total of one hundred thousand marks for the five apes of the Berlin Aquarium, the Berlin papers eagerly seized upon this piece of news. The transport of Pongo to Europe received attention also because it was the work of German Africa explorers. Soon there was talk of an "epoch-making event," and the tone became almost histrionic.

In retrospect, one would do well to keep a clear head when dealing with all of this. Whether "zoo directors and learned men from all over Germany" were already hurrying to Berlin at the beginning of July is something we simply do not know, just as we are in the dark about the background to Hagenbeck's offer. Little had leaked to the outside even about the work of the guardianship committee that was overseeing the ape's welfare. The only thing that is clear is that the rush of visitors manifestly exceeded the Aquarium's capacity.[1] Complaints proliferated during the eight hours that Pongo was on display every day. As early as 4 July, management had to appeal to the understanding of the curious throngs: because the "precious ape" needed rest during his adjustment period, it was not possible to satisfy all wishes. Anyone who was still eager to catch a glimpse of him during this time would have to be patient. Or he could stand outside the residential house of the stock corporation. For it was here, in the residence of Otto Hermes, that Pongo was housed during the midday hours and overnight. And when he appeared at the open window in the evening with Hermes or his keeper, traffic ground to a halt, and we are told that the police had to repeatedly "admonish" the crowds "to maintain order." The strange scene, which was repeated for about five weeks, was described by the writer and editor Paul Hirschfeld in the *Illustrirte Sonntagszeitung*: "It was very comical when the gorilla, around six o'clock, sat at the window of the director's apartment on the second floor and looked down upon the people gathered in a large crowd along Unter den Linden. He merrily applauded those below, and his mirth increased when the people called 'Hooray!'

in response. Since he could not join it, he gave the most varied signs of his approval."[2]

Pongo's "human-like manner" and his drolleries fascinated the visitors. The saying about "the most popular resident of the capital of the German Reich" made the rounds. Journalists were describing the gorilla as "king of the apes" and as a "dignified Grand Mogul." In the summer and fall, many German family magazines ran stories in which authors like Bruno Durigen and Heinrich Leutemann wrote about the happenings at the Aquarium. Full-page drawings with "gorilla studies" supplemented the reading. Even decades later, a contemporary witness remembered a childless couple that had traveled from southern Germany to Berlin "just to see this ape!," to the "utmost outrage" of all their cousins and aunts. For the operators of the Aquarium, the exhibit also paid off financially. Already in July, the books showed additional revenue of seventy-six hundred marks, and by the end of the year the stock corporation recorded three hundred and thirty thousand visitors, a new single-year high.[3]

THE DAILY ROUTINE

Normally, the gorilla was brought into the Aquarium after breakfast and put on display. The cycle was repeated in the afternoon, with some scientists given a private audience. In the director's apartment, too, a strictly regulated daily schedule applied to Pongo, as described by Otto Hermes:

> In the morning around the eighth hour, the gorilla rises from his bed, sits up straight, yawns, scratches a few spots on his body, and remains drowsy and listless until he has taken his morning milk, which he is wont to drink from a glass. Now quite invigorated, he leaves his bed, looks around the room to see if he can find some object for his appetite for

destruction, looks out the window, starts to clap and—in the absence of suitable company—to play with the keeper. ... At nine o'clock he is washed, which delights him, and he expresses his pleasure with a grunting noise. In keeping with their cohabitation, he takes his meal at the same time as the keeper. For breakfast he is given a pair of wieners, frankfurters, or Jauersche sausages, or an open-faced sandwich topped with smoked Hamburg meat, Berlin cow cheese [a soft cheese with caraway seeds], or something else. His favorite thing to drink with it is his Berlin White [beer]; it looks exceedingly comical when he grabs hold of the voluminous glass with his short, fat fingers, and he would drop it if he did not use one foot to assist him. He likes to eat fruit, and a lot of it, with cherries he carefully separates the pits. Around one o'clock, the wife of the innkeeper brings him his food.... Once the woman appears, he inspects the dishes and likes to nibble from what tastes best to him. A box on the ear is the usual result of his fondness for sweet things, and he then politely waits for the beginning of the meal, not taking his eye off the dishes for a second. First a cup of bouillon. In a flash he has drained it down to the last drop. Then there is rice or vegetables, mostly potatoes, carrots, or kohlrabi cooked with meat. The woman insists that he behave properly, and indeed he is already using the spoon skillfully. But as soon as he thinks nobody is watching him, he sticks his mouth right into the bowl.

Pongo in Berlin, 1876

> Finally, a piece of roasted chicken is what he likes best. He is not picky when it comes to food; what the keeper eats is also his fare, and he is not far behind in quantity. Once the meal is over, he wants some peace and quiet, like any person. A midday sleep of one to one-and-a-half hours gets him ready to play again. In the afternoon he is given fruit, in the evening milk or tea and a bread-and-butter sandwich. At nine o'clock he goes to sleep. He lies on a mattress wrapped in a wool blanket. The keeper sits by his side until he has fallen asleep, which does not take long given his great need for sleep. He prefers to sleep in the same bed as the keeper, whereby he hugs him and places his head somewhere on the keeper's body. He sleeps soundly through the night and does not usually awaken before eight o'clock.[4]

A GLASS PALACE OF HIS OWN

During the first few weeks, Pongo still had to share his cage in the Aquarium with baboons and guenons. But as early as 15 July, the *Berliner Fremdenblatt* reported on construction activity. And in fact the following days saw the beginning of demolition work at the rock grotto with the large aviaries. New quarters were being created with a separate heating and ventilation system: a spacious cage connected to a small palm house in which a water feature was burbling. A few walls of this "gorilla dwelling" were faced with mirrors, which led the press to speak a bit pompously of a "glass palace." Even though the attempt to create artificial tropical air increased the risks to Pongo's health, if anything, the project was unusual and sensational. The furnishings of the cage were a mix of living room and gymnasium. Alongside a sofa, which could be covered with a curtain if needed, there were two chairs and the

Chimpanzee, the dog Flock, and
an orangutan in the Berlin Aquarium, 1876

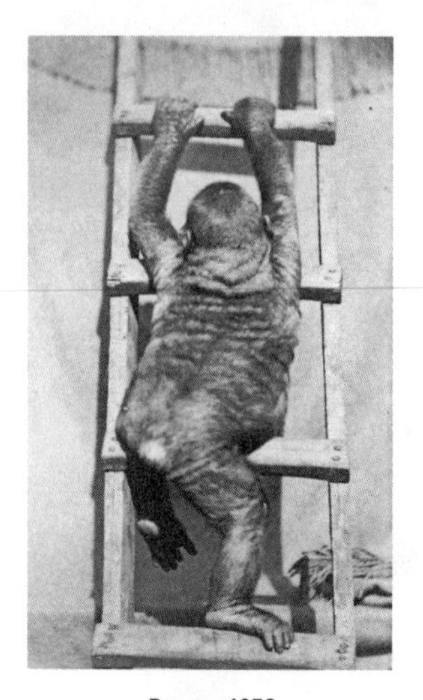

Pongo, 1876

inevitable wool blankets. Photos and drawings also reveal a ladder, two climbing poles, and several ropes.[5]

The modifications, which cost around twelve thousand marks, were completed by the end of August. From then on, Pongo spent his time separated from the other monkeys and apes. Only the female chimpanzee Tschego continued to keep him company. This was done to prevent the gorilla from being seized by "homesickness and melancholy"—a danger against which many experts warned with great seriousness. The scuffles and tussling of the young animals entertained the visitors and came to an end only when Tschego was taken ill and died from tuberculosis. After some back and forth, the gorilla was now placed into his cage together with a small dog called Flock. The dog grew up in the "glass palace," and we are told that he himself soon took on "something ape-like." Always on guard against Pongo's rough drollery, he had learned to snap at him lightning-fast and to earn his respect.

In the fall, the rush of curiosity-seekers slowly began to wane. Pongo was now taken into the office of the Aquarium to sleep, where he occupied a room with two windows to the courtyard. But ahead of Christmas, management decorated the gorilla's "glass palace" with fir trees and lowered the admission fees. For three days, visitors could now see the house and its residents at half price. It was not uncommon for visitors to turn around and spend the money they had saved on "Aquarium guides" and photographs. From the Berlin period alone, we have somewhere between four and six images, nearly all of which were taken by Adolf Halwas. Because of the long exposure times and the bulky cameras, photographing living animals was difficult. Added to this were the unfavorable light conditions in the Aquarium spaces. What the photographer accomplished is therefore all the more remarkable. His pictures resemble portraits and show an ape full of almost philosophical thoughtfulness. Only in one photograph is the gorilla climbing the ladder, which seems clumsy and touching.[6]

Pongo with a chimpanzee, 1876

Since the printing of photographs was expensive, the photographic images were often reproduced as woodcuts or lithographs. Several editions of Brehm's *Illustrated Life of Animals* include a page with gorilla scenes, all of which depict Pongo. These and other works were produced by the animal painter Gustav Mützel, who regularly visited the Aquarium and the ape. In the creative process, Mützel had to set aside his own ideas and follow the wishes of the authors and publishers. The gorilla pictures he drew adhered to the photographs down to the smallest details, regardless of whether they were from Africa or the Berlin "glass palace." A sculpture of Pongo by Friedrich Thomas strikes one as equally realistic. This was not the first time that the Berlin sculptor had worked on a commission from the stock corporation. His plaster busts of the chimpanzees and orangutans were offered for sale in several newspaper ads. Together with casts of the front and hind paws, they cost fifty marks each, a handsome sum. Potential customers were museums and scientific societies.[7] The Aquarium's financial statement of 1878 even listed a separate "plaster bust account," though the revenue was modest. In the end, it was the photographs above all that were disseminated by way of various printing techniques. They were published for decades in animal books and scientific studies, thus leaving behind a trail of their own.

CHAPTER 8

Under Observation

A "PATENT OF NOBILITY"

In the third week of September 1876, Otto Hermes traveled to Hamburg, where the Meeting of German Natural Scientists and Physicians (Versammlung deutscher Naturforscher und Ärzte) was being held. The annual gathering of the Society was one of the high points of scientific life during the empire and held out the promise of wide attention. Once again nearly a thousand members and guests had turned up, and for one week they filled the meeting space along the Elbe. There, in the "Sagebiel Establishment," the director of the Aquarium stepped to the lectern to speak "about the gorilla and his closest relatives." It was a witty account filled with stories from everyday life. In small scenes, Hermes painted an anecdotal picture and compared the behavior of all known apes. So far this kind of study had been possible only at the Berlin Aquarium, which is why the audience hung on every one of his words:

Our institution has for years placed great stock in the ownership of anthropoid apes and has kept them for a longer period of time with success and luck. Every visitor is more or less used to finding one of his cousins, a Mr. Chimpanzee or Miss Orang. Within the last year, it has managed to obtain all four anthropoids, the gibbon, the orangutan, the chimpanzee, and the gorilla.* I therefore had the best opportunity to undertake detailed studies of their life in captivity. . . . At the lowest rung is the smallest of these, the gibbon, at the same time the most delicate and dexterous of all. A face framed by a white beard and the fantastically long arms lend him a peculiar appearance. He is the only one who always has an upright gait when walking on level ground. . . . Compared to him, the orangutan is a clumsy and phlegmatic fellow. Trusting and amiable when young, he becomes increasingly wild and unruly with age. It took years and daily treats before I could hazard to approach the large orangutan, the largest ever in captivity. In his appearance he was truly beastly. The red, shaggy hair, the closely spaced, small, wily eyes in the smooth face, his disgust-evoking manners, and the horrible set of teeth made him appear like a fiendish monster, whose sight would hardly make you believe that inside of him was a nature that was on the whole sweet-tempered. It does not take much imagination to deem him a highwayman lurking in the forest.

In contrast to the clumsy orangutan, the chimpanzee offers a picture of high-spirited vivacity and dexterity, far superior to the former in intelligence. The most amiable of all the chimpanzees, Tschego, knew her surroundings precisely and obeyed every word. As a lady who cherished cleanliness

* Gibbons are today also referred to as lesser apes, while bonobos or pygmy chimpanzees, which are among the great apes, had not yet been discovered at the time.

> above everything, she cleaned and polished the glass panes of her cage. She knew and used the keys that led to the general monkey cage and even picked them out from the key ring. . . . When she succumbed to a heart ailment, I felt as though I had lost an old acquaintance. But the most noble of all is the gorilla. It is as though he has brought a patent of nobility into the world with him. Our approximately two-year-old male gorilla has reached a height of nearly three feet.** His body is covered with silky-soft, gray-flecked hair, which is reddish on the head. His sturdy, squat body, his muscular arms, his smooth, shiny black face with the well-formed ears, the large, intelligent, mischievous eye impart to him something strikingly human-like. He would resemble a Negro boy if the nose were more shapely. This impression is heightened by the clumsiness of his entire being. . . . When he, sitting there like a pagoda, lets his gaze roam across the public gazing at him in wonderment and then suddenly claps his hands while nodding his head, he has swiftly conquered the hearts of everyone. . . . He likes to interact with a large crowd, distinguishing young from old, male from female. He is kind toward children of two to three years old, he likes to kiss them, he puts up with everything, without ever making use of his superior strength.[1]

The lecture, subsequently printed, contains additional details. In plain sentences, Hermes described the games of the apes and the daily routine in the Aquarium. That Pongo would join his keeper for breakfast or "partake of a cup of bouillon" aroused astonishment. After all, it demonstrated how human-like the life of the "little anthropoid" was. Was it possible—a question posed not only by the experts—to educate the gorilla, and where, in the end, was the line

** A Prussian foot (Rhine foot) is the equivalent of about 12.3 inches.

that separated him from man? Otto Hermes was familiar with these discussions and shied away from making pronouncements. Many philosophers and scientists had already speculated about this. The zoo director Heinrich Bolau, who had dedicated a thirty-page publication about the "human-like apes of the Hamburg Museum" to the assembled scientists and physicians, also emphasized the inadequacy of what was known at the time. Even though "a small library" had appeared in recent decades about the gorilla, there was much about the rare jungle dwellers that remained contradictory and mystifying.

THE METROPOLIS OF THE SCIENCES

The Meeting of German Natural Scientists and Physicians, founded in 1822, assembled in a different location each year. Excursions and exhibits, lectures and publications were intended to convey to participants a picture of the potential and vigor of the host cities. When the congress met for about a week in Leipzig, Bonn, or Munich, it was considered an honor for the hosting scientists and physicians. But the true metropolis of the sciences was Berlin. "In no European city is the thinking bolder, more free of prejudice, and more comprehensive," noted the Danish literary critic Georg Brandes in 1879, mentioning half a dozen famous names. The institutions and collections of the university, the Charité, and the Royal Prussian Academy were located in a single building complex or only minutes apart. Their professors and assistants knew each other well beyond their work, and many belonged to several organizations at the same time. In the nineteenth century it was normal to meet with other citizens after a day's business and pursue shared goals. Scientific topics also played a role here, and organizations such as the Friends of Science in Berlin (Naturforschende Freunde zu Berlin) or the legendary Monday Club invited their members at regular intervals to lectures and an exchange of their experiences.

In November 1869, half a year after the opening of the Aquarium "Unter den Linden," the Berlin Society for Anthropology, Ethnology, and Prehistory (Berliner Gesellschaft für Anthropologie, Ethnologie und Urgeschichte) was founded in the Mohrenstraße. Many of the actors around Pongo were members: Rudolf Virchow, Robert Hartmann, Adolf Bastian, Heinrich Bodinus, Otto Hermes, and, later, Julius Falkenstein. The tasks it set for itself were wide-ranging. Using scientific methods, such as the measurement of bones and skull shapes, scientists wanted to discern the commonalities and differences of the "human races." It went without saying that the "anthropomorphic apes" played an important role as reference material. While experts were still skeptical about Charles Darwin's theory, a few of his followers were drawing far more radical conclusions. Under these circumstances nothing made more sense than to gather facts and discuss the findings. The ape question took up a lot of space in the *Zeitschrift für Ethnologie*, a journal of the newly created Society, and as early as the summer of 1876 the first drawings with the image of Pongo were printed here.

Robert Hartmann in Africa, ca. 1860

FIRST FINDINGS

Those doing research on apes included the ethnologist, zoologist, and physician Robert Hartmann. He taught at Berlin University and had spent years in northeastern Africa. In a picture of his study (page 89), we can make out a human skeleton and two ape busts. Lying on the small wooden table in the foreground is the body of Molly from the Berlin Aquarium. The female chimpanzee appears to have been posed for the photograph. Carefully framed by small corals, the body gives the impression of something precious. Indeed, the cadavers of "anthropoid apes" were something special, and the small number of experts fought over the privilege of examining them. Hartmann had already published several articles on the topic, and, as a board member of the Society for the Exploration of Equatorial Africa, he had been in contact with Falkenstein and the other Africa explorers. He received the chimpanzee and gorilla specimens sent by the expedition to Europe and dissected them at the Anatomical Institute.

Hartmann had also reported early on about Pongo and in 1876 had prepared the public for the ape's possible arrival in Germany. The fact that most zoologists at the time were doubtful about the value of studies on live animals did not bother him. Trained in field research through his expeditions to Africa, he knew of the importance of observations. While Rudolf Virchow, the famous pathologist, brought his entire family for a visit to the Aquarium and looked at Pongo "more with childlike than with learned or philosophical pleasure," for Hartmann the gorilla was from the beginning an object of study. He carefully continued the body measurements he had begun and noted the ape's weight and the width of his nose. Pongo played an important role in his books about apes that were published over the next few years and translated into several languages.[2] Hartmann described how the gorilla's skin, initially afflicted with scabies, healed and how the fur changed its color. He explained when Pongo bared his teeth, how often he pushed the small red tip of his tongue to the outside, and that the typical "neck hump" could already be discerned in young animals. With the eye of a gifted draftsman, he sketched the eye of the ape and described it in precise detail.

Robert Hartmann was not the only one pursuing scientific goals at the Aquarium. In January 1877, Baron Johannes von Fischer of Gotha announced his arrival. The future zoo director of Düsseldorf, whom Darwin, owing to an article on the sexual behavior of apes, called a "careful and acute observer," was given a private audience and was allowed to study the gorilla in peace and quiet. In the process he showed a particular interest in the expressive abilities of the ape left alone in his presence:

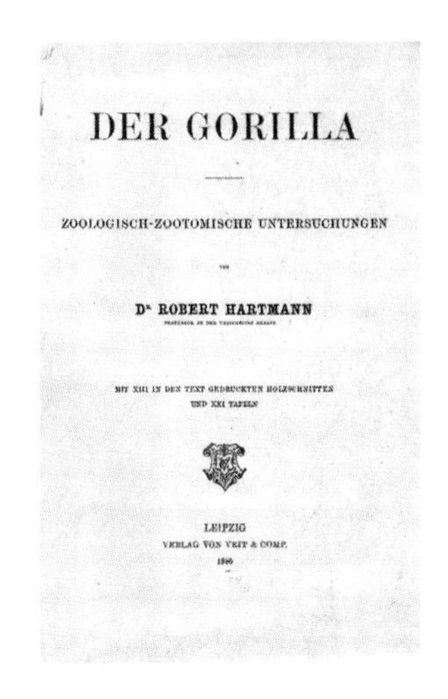

DER GORILLA

ZOOLOGISCH-ZOOTOMISCHE UNTERSUCHUNGEN

DR ROBERT HARTMANN

MIT XIII IN DEN TEXT GEDRUCKTEN HOLZSCHNITTEN
UND XXI TAFELN

LEIPZIG
VERLAG VON VEIT & COMP.
1880

> I sat down on the bed, took him into my lap, and then used everything I knew about various ape languages and dialects, along with the obligatory facial expressions, to get him to converse and be able to study his face. In vain. A macaque or a baboon would have understood me right away and would have answered like with like. Pongo did not understand me. He looked at me puzzled and seemed to see in my efforts only meaningless mumbling and grimacing, without so much as uttering a sound. Finally he slapped me with a grin and right after that bit my nose.[3]

Since Pongo was getting more and more carried away in the game, the baron eventually had trouble asserting himself over the gorilla. At least he registered that the gorilla was able to giggle like a human and was evidently ticklish. As was done with nearly all apes, a mirror was held up to Pongo to see if he would recognize his reflection. Four months later, the Leipzig physician Arthur Kollmann had different objectives. The young man needed unusual material for his doctorate. His thesis, titled "A Comparison of the Human and Ape Hands," contained an "imprint of the tactile elevation of the first order" that came from the "Berlin gorilla." In the text, Kollmann thanked Hermes for the opportunity to repeatedly observe "the small, trusting fellow" and presented his very specialized findings.

We do not know whether there were other requests for consulting hours in the "glass palace." Only the names of a few scientists who visited the gorilla can be ascertained. For example, the zoology professor Wilhelm Peters came from Berlin, and Karl Freytag, who did research on domestic animals, traveled from Halle. All of them were also intrigued by the question of how much Pongo would develop over the course of his life. For as much affection as he was garnering now, the "darling of the Berliners" was under observation. After all, the experts were assuming that the gorilla's aggressiveness and unpredictability would increase with age and that his brain would even shrink then. While the young animal was considered human-like, the fully grown specimen was seen as a dangerous beast. That alone can explain the curiosity that seized many writers when Pongo bared his teeth or tested his strength against the iron bars of the cage. For now, though, the scientists were content to make suggestive comments and to wait and see. And as long as the little ape romped through the cage with the dog Flock and engaged in his hijinks in "divine gorilla mood," he was captivating the hearts of both visitors and specialists.

ON THE SICKBED

One source of recurring excitement after Pongo's arrival was his illnesses. Instantly the normal workday came to a halt and nervousness pervaded the Aquarium. The very first time it happened, a "health council" of four professors convened, and Julius Falkenstein hastened back to Berlin from the Prussian provinces.[4] On 8 September 1876, the gorilla was like a completely different animal and was tossing around "coughing and wheezing." Hermes had a sickroom set up for the ape in the "reception room of his office," in the middle of which stood a makeshift "four-poster bed." A white muslin cloth was stretched over the legs of an upturned table, and the gorilla lay under it on top of soft blankets. Falkenstein took over the treatment together with the physician Martini. The two men visited the patient four times a day. As it turned out, Pongo was suffering an "inflammation of the trachea," which was treated with quinine and Emser Kränchen (water from the Bad Ems spa). Since many papers were reporting on the events, the public was very invested emotionally. After Otto Hermes published his account, "more than one hundred inquiries" came in every day of people seeking more information. The worst was over by 17 September, and a week later the *Berliner Fremdenblatt* was able to give the all-clear: "Today we can report the splendid news from the Aquarium that friend gorilla is completely well again and has left his sickbed wide awake, in other words, that the terrible, gorilla-less time is over."

Pongo, 1875

But in the end things were not that simple. Pongo fell ill again in November, this time with an inflammation in his mouth and pharynx. Against the advice of the professors who had been summoned,

Hermes decided to treat the ape with silver nitrate, and after administering it repeatedly he was successful. In the process he had three keepers hold down Pongo while he personally carried out the "painful procedure." The director of the Aquarium followed his intuition. At the same time, the event shows just how unsure the experts still were. Fundamentally, contemporaries knew hardly anything about the needs of an ape and treated Pongo no differently than a human child. That is hardly surprising in an age that knew neither the causes of nor the appropriate treatments for many illnesses (for example, tuberculosis). Although the Frankfurt zoo director and veterinarian Max Schmidt had published the first book about the "diseases of apes" in 1870, he could do no more than summarize the existing knowledge. But at least the book contained autopsy reports on fourteen apes, which were helpful in describing certain hazards. Additional reports were published by the Society of the Friends of Science in Berlin in its proceedings. Anyone who reads them is confronted by the suffering of the animals and the helplessness of their keepers. These are devastating accounts that become engraved in one's memory. Some of the chimpanzees and orangutans progressively lost their teeth; others wasted away while suffering cramps.[5] The history of imported apes is a history of loss. Most of the "anthropoids brought to Europe at great effort" died after just a few months or even weeks. Only in exceptional cases can one identify an animal that survived several years. Added to this is the fact that most animals never even reached the continent and perished in agony in their countries of origin or during transport.

There was virtually no criticism of these conditions, which persisted well into the twentieth century. Only once, in 1881, did the journal *Der Zoologische Garten* (The Zoological Garden) print a piece in which an author we can no longer identify wrote very candidly, "It must fill an animal lover with pain when he sees how every year, with great sacrifices and at enormous cost, new specimens of the anthropoid apes are dragged north, apes that likely never remained alive for long here. These animals are thus knowingly consigned to death, and it does not make any difference that every time one hears the claim again: 'Now the conditions that are necessary to preserve their lives here have been discovered and put in place.'"[6] Otto Hermes, for one, knew about the dangers to which Pongo was being exposed, and he strove to keep them to a minimum. Every day the ape's room was disinfected with calomel, a mercury compound, and the temperatures in the rooms where he spent his time were not as high as in other zoos. Julius Falkenstein likely gave the director of the Aquarium precise instructions in this regard and warned against overdoing it. But Pongo's diet, which included fatty and meat-containing foods alongside beer and wine, in no way accorded with his needs. As of yet the experts had no clue that the gorilla, whom they described as an omnivore, was decidedly a vegetarian.

CHAPTER 9

"The Only Gorilla Is Coming"

A TRAVELOGUE

In April 1877, the Berlin correspondent of the *British Medical Journal* was granted a "private audience" with Pongo. Impressed, he composed a lengthy report that was reprinted in numerous papers. It was not the first time that the English press took an interest in the ape, though this time there was an additional reason. There was a growing number of indications that the gorilla was going to be sent on a tour. Offers were on the table from London, Vienna, and Paris, and over the course of the summer an agent from New York also made contact. Pongo's "guardianship committee" initially recommended a stay at the Thames, on the grounds, we are told, that the "damp climate" would have a beneficial effect on the animal's well-being. In reality, though, it was about money. Large sums were being offered by the Royal Aquarium in London, where the exhibition was to take place. Since Pongo fell ill again in May 1877 and struggled for weeks with an intestinal catarrh, management put the trip on hold. It was not until 15 July that the small retinue set out. In addition to Otto Hermes, it included the keeper Wilhelm Viereck and another staff member. The gorilla was also accompanied by the chimpanzee Tschego and the dog Flock, to whom Pongo had become accustomed. Five months later, Hermes delivered a public lecture about his experiences in the British metropolis. It was a look back full of emotion and pathos-laden descriptions. The recounting of the trip over the English Channel, which took place in stormy weather, must have astonished many listeners:

> On 17 July, we left Hamburg on the ship *Castor*. The only passenger who did not get seasick was Pongo, and when, in the vicinity of the Dutch coast, one after another of his fellow human travelers, whose darling and even table companion he had instantly become, withdrew quietly to his cabin or into a quiet corner, he was seized by a great melancholy on account of this. With a concerned and sad look he went from one to the next, placed his hand on their head, and looked at them with a gaze that attested to his genuine concern and seemed to be constantly asking: "What's wrong with you?" As long as there was still life on the ship and the temperature was mild, he frolicked on board and in the masts in high spirits, and no one had any worries about him on account of this, for it was known how exceedingly careful he was. In London, where some passengers had quite a hard time parting from him, the party initially put up in the hotel, but already the following day, 19 July, he was able to switch to his residence in the local aquarium, which was quite comfortable for an ape. Here he was first presented to the representatives of science and the press, but then put on display for the public for three hours a day. The preparations and the entire manner of the exhibition

had been carried out in true London fashion, that is to say, with an apparatus of advertising and pomposity that would be entirely unthinkable here in Germany, and, most of all, is probably quite illegal. For six weeks prior to Pongo's arrival, the streets, newspapers, and trains had already been flooded with announcements, and in the aquarium itself a giant poster thirty feet long and twenty feet wide alerted visitors to the wonder animal and announced that Pongo (and that is the only way he was talked about) would be receiving "his visitors" at some hour or another. The entire aquarium, which in London holds close to ten thousand people and is used for all kinds of exhibitions, and is thus even much less so than the Berlin institution merely an aquarium in the truest sense, was constantly overcrowded and all were pushing for one of the only three hundred seats (the Englishman will watch something only seated) in the special room in which Pongo was holding his "consultation hours." The gorilla had such magnetic power over London's population that he completely eclipsed the throngs coming to see thc talks about humbug given at the same time and on the same premises by the much-heralded American Barnum.* All papers, chief among them the *Times*, published multicolumn articles about Pongo every day. The royal family, all the ministers, and dignitaries came to see him—in short, the gorilla ruled all of London.[1]

IN THE ROYAL AQUARIUM

The article that resulted from the director's lecture did not appear until the spring of 1878. At that time, the events surrounding Pongo

* Here "humbug" refers to circus attractions. Because of his sensational projects, Barnum was sometimes also called the "Prince of Humbugs."

were already several months in the past, and perhaps that is why Otto Hermes had romanticized certain elements. The director was impressed by the dynamism of the city of four million, and since he spoke English poorly, he did not always understand what was being related in London. But Hermes was on the mark when it came to his assessment of the Royal Aquarium. It was hard to compare the sprawling structure across from Westminster Abbey with the Berlin institution "Unter den Linden." Even if the man responsible for the aquarium section was William Lloyd, the fish and aquatic animals were at best secondary here. Rather, the event palace, which had opened its doors on 22 January 1876, was one of the establishments in which the well-off residents of the city amused themselves. Just the main hall with its domed glass roof was more than one hundred meters long and nearly fifty meters wide. Furnished with basins and sculptures, fountains, and a skating rink, it offered space for many events. Those for whom this was not enough could entertain themselves in billiard and reading rooms, galleries and exhibit spaces, or could attend the concerts of the Aquarium's own orchestra. Nearly three hundred full-time employees and seven hundred seasonal workers toiled in the building, and yet its owner was in trouble. Just one year after the inauguration of the facility, signs were pointing to a disaster, with the deficit rising to twenty thousand pounds. High time to procure new attractions and tout them to the Londoners.

Pongo was not the only sensation used to woo visitors. "Miss Zazel," who had herself shot through the air like a human cannonball, was performing at the same time, while only a few steps away the American circus pioneer Phineas Taylor Barnum was telling stories from his eventful life. Concerts and festivities alternated, and the building's theater, inaugurated three months earlier, offered an extensive program. But a look at the papers shows that Otto Hermes and his gorilla still managed to cause a stir. The *London Times* announced Pongo's arrival on the front page, and over the

The Royal Aquarium in Westminster, ca. 1876

following weeks it published several additional reports about the ape. Other papers also dispatched their reporters to the Aquarium.[2] Much of what the journalists noted down is already familiar to us: Pongo was now a "very nice beast," "the lion of the season," or simply "the most wonderful monkey I have ever seen." Added to this was the attention from around the world, since news was now spreading across the ocean faster than ever before. The *New York Times,* the *Melbourne Telegraph,* papers from Australia and New Zealand—they all let their correspondents send reports or reprinted the English press. The same is true for Paris, where the *Journal des Voyages* even published an illustration. And already back then, some pieces of information were lost on the way to the editorial offices or changed to the point of being unrecognizable. For example, "Dr. Falkenstein" suddenly turned into a "Dr. Frankenstein," and the German expedition to explore equatorial Africa became "the famous Prussian expedition" that had brought the gorilla to Berlin.

On the floor plan of the Royal Aquarium one can identify the place where Pongo could be visited. It was a room in the northern section of the building. The cage erected here consisted of thin, relatively widely spaced iron bars and stood slightly elevated on a platform. Several ropes hung from the ceiling; a ladder and two armchairs completed the furnishings. Once the visitors had sat down, the gorilla was brought in along with his entourage (Flock the dog and Tschego the chimpanzee) and left to himself for the first few minutes. But then the keepers swung into action, and, as was done in shows offered by traveling menagerie owners, they delivered a short, memorized presentation. The curious onlookers did not learn much, for the vocabulary of the two Germans was only enough for the bare essentials. Finally, to justify the steep admission fee of three shillings, Pongo was taken out of the cage and placed on a chair in the midst of the audience. The subsequent "escape of the ape" regularly caused excitement and offered those present a chance to experience the gorilla in action. This spectacle was repeated three times a day, and anyone wishing a "private audience" in between had to pay five shillings.

In contrast to the presentation "Unter den Linden," this sequence of events was reminiscent of a vaudeville or circus operation, with the financial interests of the organizers dominating everything that happened. The sale of photographs and plaster busts, some of which were produced only in London, was intended above all to boost revenue. Thus, the well-known taxidermist Rowland Ward modeled a second sculpture of Pongo, and additional gorilla photos were taken in the studio of the London Stereoscopic Company. Those evidently included also the photograph that shows Pongo and the keeper Wilhelm Viereck. Embracing tightly, the twenty-seven-year-old Viereck and his ward look into the camera. It is the only image we have of Viereck, and it is also highly valuable because it underscores the close relationship between the two. No one spent more time with the gorilla and cared more intensively for him than the aquarium keeper from Berlin.

DARWIN'S SILENCE

The extensive reporting, the numerous ads in the *London Times*, and the poster ads helped make Pongo's visit a complete success. When twenty-five thousand visitors were counted at the Royal Aquarium on especially crowded days, the seats in front of the gorilla cage were sufficient for only a fraction of the onlookers. The visitors included several members of the English royal family and the Prussian crown prince Wilhelm, the prime minister, and members of Parliament. Most of all, however, the old guard of British scientists showed up. On 21 July 1877, two days after Pongo's arrival, Richard Owen came to inspect the gorilla. The still sprightly Owen held great authority and was working at that time on setting up the Museum of Natural History. The zoologist St. George Mivart and the surgeon and journalist Frank Buckland had joined him in accepting the invitation to the Aquarium. The three men, some of whom had spent years on the comparative anatomy of apes, rejected the Darwinian theory of evolution, and the sight of Pongo seemed to confirm their opinion. If the newspaper reports can be trusted, Buckland tried to teach the gorilla how to write, but all he accomplished was that Pongo put the approximately three-centimeter-long pencil into his mouth and swallowed it.

This minor incident, which also included news that the ape had then drunk a glass of beer and devoured a slice of roast beef, was also circulated because the press was hoping to provoke a reaction from Charles Darwin. Unlike in Berlin, in London hardly a journalist failed to point out the importance of the gorilla for the theory of evolution. But Darwin was holding back and avoided a public statement. As far as we know, he remained silent during the eight weeks during which Pongo was exhibited at the Thames. After he had completed his book *The Expression of the Emotions in Man and Animals,* the anthropoid apes occupied him only marginally, his interest being devoted to flower biology and carnivorous plants.

Pongo with the keeper Wilhelm Viereck

All the more active was Frank Buckland, who examined and then described the gorilla repeatedly. Unlike Darwin, the eccentric book author enjoyed the attention of the public and quickly got carried away into staging some unusual presentations. His posthumously published *Notes and Jottings from Animal Life,* which contains a separate chapter on "Mr. Pongo," includes the following observation:

> A little boy and girl came in to see him while I was present. After a while they both began to play with Pongo. Gradually they fraternised, and began to play together after the manner of little children. Not being a child, I cannot enter into their funny saying and doings about nothing at all. So these three, the little boy and girl and the gorilla, played together after their own childish fashion for nearly half an hour, and I made the children experiment on him with ornaments, handkerchiefs, &c.; but no—the ape's brain could not understand the human. Pongo put everything in his mouth, and tried to bite it up.

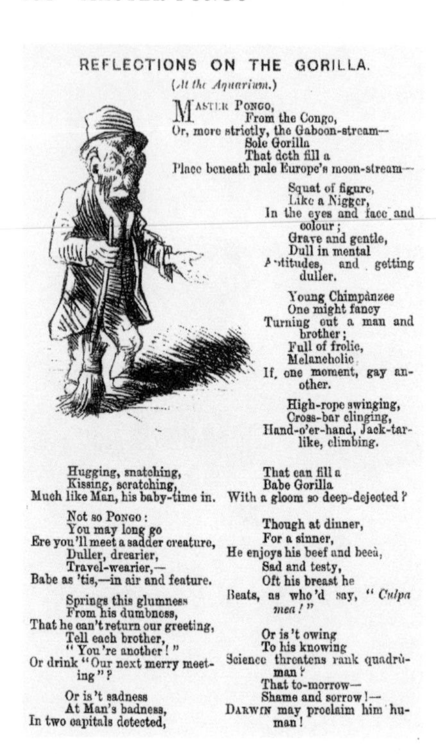
REFLECTIONS ON THE GORILLA.

(*At the Aquarium.*)

MASTER PONGO,
From the Congo,
Or, more strictly, the Gaboon-stream—
Sole Gorilla
That doth fill a
Place beneath pale Europe's moon-stream—

Squat of figure,
Like a Nigger,
In the eyes and face and colour;
Grave and gentle,
Dull in mental
Aptitudes, and getting duller.

Young Chimpànzee
One might fancy
Turning out a man and brother;
Full of frolic,
Melancholic
If, one moment, gay another.

High-rope swinging,
Cross-bar clinging,
Hand-o'er-hand, Jack-tar-like, climbing.
Hugging, snatching,
Kissing, scratching,
Much like Man, his baby-time in.

Not so PONGO:
You may long go
Ere you'll meet a sadder creature,
Duller, drearier,
Travel-wearier,—
Babe as 'tis,—in air and feature.

Springs this glumness
From his dumbness,
That he can't return our greeting,
Tell each brother,
"You're another!"
Or drink "Our next merry meeting"?

Or is't sadness
At Man's badness,
In two capitals detected,
That can fill a
Babe Gorilla
With a gloom so deep-dejected?

Though at dinner,
For a sinner,
He enjoys his beef and beeà,
Sad and testy,
Oft his breast he
Beats, as who'd say, "*Culpa mea!*"

Or is't owing
To his knowing
Science threatens rank quadrùman?
That to-morrow—
Shame and sorrow!—
DARWIN may proclaim him human!

Poem in *Punch*, August 1877

> When the two humans and the gorilla were sitting at play on the floor I could not help seeing the amazing difference between the countenances of the gorilla and of the children; the one decidedly and purely monkey, the others decidedly human. I could not in fact help seeing what a vast line the Creator had drawn between a man and a monkey.[3]

Frank Buckland's intent is as clear as it was futile, since the popular ideas associated with Darwin's name had long been shared by many contemporaries. As though it were perfectly natural, most journalists highlighted the "human-like nature of the gorilla," and visitors arrived at similar conclusions. For weeks, the influential satirical magazine *Punch* followed the exhibit in the summer of 1877 and made its own sense of the whole thing in the form of a poem. Sixteen years earlier the editors had published a series of widely noticed gorilla illustrations, and now the journal returned to the topic. Two drawings depicted Pongo with hat and coat being interviewed. The reporter, claiming to have discovered that the gorilla was to be elected secretary and treasurer of the Aquarium company, posed a number of questions that quickly drove the game of deliberate confusion to the level of the absurd. And so the "learned ape" went on to pronounce what he thought of William Shakespeare and why he had a special affinity for the libretto of Richard Wagner's *Rheingold*. The "interview," laced with political and artistic potshots, was not the only attack launched by *Punch*. In the so-called Pongo-isms, short commentaries on current affairs, all sorts of irreverent words were put into the gorilla's mouth. A poem printed by the magazine shows that a good deal of sympathy for the young ape stood behind the biting commentaries.[4] As so often, the artists were the first to set a new tone and give voice to a widespread sentiment.

CHAPTER 10

Final Certainty

A CATASTROPHE

Pongo's trip to London was also closely followed in Berlin. Short articles appeared at regular intervals, and the capital's press reported on the events. The topic of the gorilla fascinated many. At the end of June, Baron Hugo von Koppenfels had published a lengthy article in the *Gartenlaube* about his hunt for the "mysterious apes," which had aroused curiosity. The "original drawing" published for that occasion depicted a giant gorilla in attack posture. Such images are difficult to associate with Pongo, who was "celebrating triumphs" in the British metropolis. The ape was and remained the darling of the Berliners, an "enchanting creature of delightful manners." No wonder, then, that there was also interest in his return from London. As the *Fremdenblatt* reported, the departure had to be postponed by several days. A storm over the channel forced Otto Hermes to extend his stay. The weather calmed down on 11 September, and around noon the gorilla was able to scramble around on the deck of the *Castor* in sunshine and at a temperature of seventy degrees. Following its arrival in the Hanseatic city, the convoy stopped over at the Hamburg Zoo, where about forty thousand visitors came to see Pongo within nine days. Although that number fell short of expectations, it was no cause for gloomy thoughts. The trip as a whole was considered a grand success: "The gorilla," wrote the *Tribüne*, "has returned healthy and sprightly, and since Monday the 24th of September he has been on display again in the Aquarium. We hardly need to mention that he caused an unusual stir in London."

No one, least of all Otto Hermes, had any clue at that moment that there was very little time left for the Berliners to see the ape once again. To the contrary, his health was "better than it has been in a long time," and already there were thoughts about further trips. In October 1877, the Norddeutsche Buchdruckerei und Verlagsanstalt published a sixteen-page booklet in which one could read the story of Pongo.[1] Reports in several dailies invigorated the discussion and attracted new visitors. However, on 13 November a shocking announcement made the rounds: around four-thirty in the morning, the gorilla had died completely unexpectedly after a "sudden death rattle and with a loud outcry." Otto Hermes, who had been summoned, stood crestfallen by his charge's bed and struggled to keep his composure. The evening before, he had examined Pongo and taken his temperature. But the diarrhea from which the ape had been suffering for fourteen days had not seemed a cause for serious concern compared to the illnesses he had survived. The loss affected the director of the Aquarium in every respect, emotionally as well as financially. In the days that followed, the number of visitors declined, and the stock price of the stock corporation dropped 3 percent.

Nearly all the papers in the capital picked up the report of the gorilla's death and printed a text that was largely identical. In addition, some journalists pointed to the animal's life span. At one year and nineteen and a half weeks, the ape had spent "a remarkably long time" in Europe, wrote the editors of the *Berliner Fremdenblatt*, who acknowledged the services of Otto Hermes and the "special keepers." The tragedy was commented upon in *Nature* and other specialized journals (such as the *Zoologische Garten*), as well as in the *New York Times* and *Punch*. However, initially there was little to report. Compared to the final hours of other apes, the accounts of which resembled the climax of sentimental novels, the gorilla's death hardly offered material for dramatizations. Only a dissection of his cadaver could provide insight into the "cause of death," and the fact that the autopsy was performed very quickly reveals how eager those involved were to find answers.

THE HOUR OF THE SCIENTISTS

"On 14 November 1877, at eleven-thirty in the morning, a small scientific cortege has assembled in the director's office at the Royal Anatomy, to pay their last respects to the body of Pongo, the great anthropoid that the members of the German expedition to the coast of Loango brought to Berlin a year-and-a-half ago": thus began the autopsy report, which was published at the end of November in the *Berliner Klinische Wochenschrift*. It was an impressive group of at least eleven men who gathered around the small body of the ape. At their head was Rudolf Virchow, the world-renowned physician and pathologist. He had just published a book titled *Die Sections-Technik im Leichenhause des*

Rudolf Virchow

Dissection room at the Institute of Pathology of the Berlin Charité

Charité-Krankenhauses (The technique of necropsy in the morgue of the Charité Hospital) and had once again demonstrated his outstanding knowledge in this field. In addition to the professors Rudolf Hartmann and Bogislaus Reichert, he was assisted by more specialists. Otto Hermes, Julius Falkenstein, Carl Nissle, Max Boehr, the assistant physicians Gustav Broesike and Hermann Rabl-Rückhard, and the artists Paul and Franz Meyerheim rounded out the group. At least three of the men were directly involved in the autopsy, with Gustav Broesike bearing the brunt of the work. The procedure was tried and tested, and by the early hours of the morning "the morgue assistant [*Leichendiener*] at the Anatomical Museum, Jean Wickersheimer, had precisely recorded the measurements of all body dimensions and had made plaster casts of the skull, the bust, as well as all extremities." They were later replicated at the university and sold by the stock corporation of the Aquarium.

As there were plenty of skeletons and taxidermied gorilla hides in German museums after the Loango expedition, scientists were interested above all in the ape's inner organs. "We found the skull already sawed open and the brain set aside for more careful examinations in a hardening fluid. A look into the exposed base of the ape's skull offered every physician the deceptively similar picture of the opened skull of a child," we read in Max Boehr's account. Boehr, a medical counselor and "physician of the District of Nieder-Barnim," not only described the opening of the abdominal cavity and the extraction of the bowels, but also addressed the search for the cause of death:

> Anyone who has stood by the countless corpses of human children who perished from diarrhea and intestinal catarrh, had to be surprised by the striking similarity, indeed sameness, of the anatomical findings of the intestinal mucosa and their content that the autopsy knife revealed in this ape child; the young Pongo died of nothing other than a simple diarrhea and intestinal catarrh.—Only a bent pin and a glove button in the appendix, foreign objects which, incidentally, had not caused any irritation to the mucous membrane, reminded the visitor that he was looking at the bowels of a beast.[2]

THE IMAGE OF THE GORILLA

Pongo supposedly died of diarrhea and inflammation of the bowels. The news spread quickly, with the papers' editorial boards commenting above all on the small objects that had been found in the body of the dead gorilla. However, five weeks later the report had to be corrected. Gustav Broesike published a very different version of the ape's last days in the *Sitzungsberichte der Gesellschaft Naturforschender Freunde zu Berlin* (Minutes of the Meetings of the Society of the Friends of Science in Berlin). Now it was a "pronounced lung infection," the dreaded consumption, that had "carried off" Pongo. Over six pages, Broesike laid out the details and dispelled any possible doubts. Since during the last weeks before the gorilla's death his temperature had been measured only during the day, the nighttime fever spells had not been noticed and had eventually led to his death.

The surprising findings now interested only the experts. The topic had already disappeared from the papers and magazines, and the excitement surrounding Pongo had died down. And yet there was a minor epilogue, for this story does not relinquish its hold on us quite yet. Much could still be related: about the new apes of the Aquarium, for example, or the growing competition from the Zoological Garden, the meetings of the anthropologists, or Julius Falkenstein, who married his lovely fiancée in May 1880. In the end, though, it is a different picture that captures our attention and ties everything together. Of all places, the second edition of *Brehm's Life of Animals*, the world-famous multivolume work by the founder of the Aquarium, published a woodcut whose depiction brings together all of the contradictions.[3] Like many of the drawings, the image was the work of Gustav Mützel, who had often observed and portrayed Pongo. This time, however, he combined several existing depictions. While we see in the foreground the sleeping ape from the first days in Chinchoxo, behind him sits the massive gorilla from Du Chaillu's descriptions. It is a scene that, in its contradictions, would shape the perception of the public for decades. The small young animal and the image of him as an adult, distorted into something monstrous, now existed side by side. The fact that this illustration appeared also in the next edition, in color no less, was probably more than a coincidence, since the new volumes were adapted by Eduard Pechuël-Loesche. Falkenstein's old travel companion, a member of the German expedition to explore equatorial Africa, thus made sure that Pongo lived on in image and text and that readers would not forget him (see illustration on page 110).

In the zoological gardens, however, gorillas remained for now a rare trophy. Keeping a gorilla was considered difficult, and for the most part the animals died within a short time. As late as 1906, a Frankfurt zoologist reported that it was impossible to display them for any length of time. The gorilla Bobby, whom the Berlin Zoo acquired in 1928, became famous for the simple reason alone that he survived seven years in captivity. The study of gorillas, too, advanced but slowly. Not until the second half of the twentieth century did the public find out more precise information about the natural behavior of these animals from the field studies of George Schaller and Dian Fossey. Their complicated group structures, how they communicated or learned—all of this was largely unknown until then. Today, when it is almost too late and the number of gorillas is declining dramatically, entire research groups are devoted to them. And so the story of Pongo in the end also reveals how long it has taken us to better understand these jungle dwellers discovered at such a late date, and to see them for what they are: our gentle and fascinating relatives.

AFTERWORD: THE "GIANT APES"

As with so many interesting topics, I came across the events surrounding Pongo by chance, even though the fate of apes brought to Europe had been occupying me for years. It all began in early 2002 at the Leipzig Zoo. Together with Jörg Junhold, the director of the zoo, I was working on a publication for the zoo's jubilee. Although the institution had its own archive, it was rarely used and was spread among various rooms. For that reason, some of the documents had to be laboriously retrieved, so while I was already working on the text, new documents were constantly arriving. Among them was a small poster with an orangutan. Standing menacingly upright, he looks straight at the observer, his open mouth baring a number of terrifying teeth. The depicted specimen was one of a group of four adult apes that the zoo's founder, Ernst Pinkert, had exhibited in Europe after 1893. The orangutan's size exceeded the wildest expectations. That such impressive creatures existed at all was known in the nineteenth century at most from travelogues. Wherever they were shown, the curious came in large crowds—hundreds of thousands in Paris alone. That was precisely what the director of the Leipzig Zoo had banked on, since such a risky investment was worthwhile only under certain conditions. The first two specimens had already cost ten thousand marks each: a lot of money, which had to be earned back. Since I was fascinated by Ernest Pinkert, I placed his business acumen in the forefront of the account. Only later did it become clear to me that there was another, far more radical view of the events at that time.

Leafing through the old newspaper reports, I was struck by the word choice of the commentators. The orangutans were not only described and portrayed, but also demonized. Here we are told of "eerie and malicious creatures," there of "sinister figures from the gloom of the jungle." And what was one to think when the author of a *Naturwissenschaftliche Wochenschrift* (scientific weekly) spoke of "gigantic monsters" who, in their "truly uncanny size and massiveness," would have struck terror into every last visitor? Ernst Pinkert, I knew, had acquired the apes from ship officers and middlemen with whom he had been in touch for some time. The first apes were brought on land in Antwerp and exhibited in Brussels only two days later. The next purchases came off much the same way. The zoo director hastened to Liverpool and Hamburg, where he received the animals from steamers and put them on display as quickly as possible. Everyone involved knew full well that the orangutans would survive for only a short time; the likelihood of being able to show them longer than half a year was low.

And in fact, the animals transferred from Brussels to Paris died within three weeks. From the Gare du Nord, they had been driven through the city's dark streets in an open car on 28 December 1893. I tried to imagine the scene: the large transport cages (which had been built back in Leipzig), the horse-drawn carts, the astonishment of the passersby, the shadowy bodies behind the iron bars. Over the next few days, all of Paris spoke about the "Pinkertean orangutans," and visitors crowded into the New Gallery of the Jardin d'Acclimatation. By that time, the two fully grown males already seemed to be completely spent and were displaying all the signs of pneumonia. Still, for the Leipzig zoo director—who had named the animals Max and Moritz after Wilhelm Busch's children's book—the excursion into the French capital proved good business. Even the cadavers of the apes were sold for a pretty sum and served the scientists for purposes of research.

As the exhibits had begun, so they ended: the star of the last orangutan group, the "man-sized Jumbo," died in 1895. His body was dissected, and the findings of the examinations filled the pages of several professional journals. For me, however, Jumbo held a different kind of importance. During his lifetime he had his portrait drawn, and later he was also sculpted. The wildly animated sculpture, a work by August Gaul, a sculptor and friend of Liebermann, is today in Leipzig. There, in the Museum of Fine Arts, I had made it a habit to pay a visit to "my apes" from time to time. And when I stood in the freely accessible and much-too-large room, I almost always felt the need to describe the fate of the orangutan one more time. Who had set the story in motion, and what had driven it? How many orangutans had already died during the hunt and the passage? And how could I talk about it today, when the survival of the apes is so starkly in question?

Eventually I came across an article from the magazine *Der Zoologische Garten*. In it, the director of the Berlin Aquarium described the journey of a small gorilla across the English Channel. It was the very report about the "triumphs" of Pongo in London. Although I considered the account exaggerated, my curiosity had been aroused. Apart from the literary quality of some texts, I was touched above all by the empathy that had been extended to the gorilla. Whether in Africa or Europe, he had always been protected and sheltered as best as the humans knew how. His case, too—there was no denying it—revolved around money and scientific ambition, control and possession. But while I had found empathy lacking in the treatment meted out to the "giant Pinkertean apes," it was shown to the little gorilla in rich measure. Could his biography offer new access to a topic I had already set aside? Was it possible to write about apes, after all?

In April 2010, I stood in a side entrance to the Museum of Natural History in Berlin. A little later, a young assistant, Saskia Jancke, led me through a labyrinth of rooms and corridors, until we came to a stop in front of a gray steel cabinet. On the far right, on the lowest shelf of the cabinet, I saw Pongo's skeleton. It was tiny, barely larger than the boxes full of bones next to it. As though frozen by the sudden light, the bony gorilla was sitting in his corner and was light in my hands when I eventually took him out. Attached to his skull was a sign with the year 1876 and the name Julius Falkenstein. The skeleton must have come into the Museum of Natural History (opened in 1889) along with the other exhibits from the university's zoological collection, and then it sat in its dark corner for decades. It was confusing that I now held it in my hands. Whatever I had learned over the previous months about Pongo and the German expedition to the west coast of Africa, about the Aquarium "Unter den Linden" and the gorilla's trip to England—it was only at that moment that the wealth of literature seemed to coalesce into a story.

SHORT BIOGRAPHIES

Barnum, Phineas Taylor (1810–1891): famous American showman and circus pioneer who also became involved in politics.

Bartlett, Abraham Dee (1812–1897): English zoologist and taxidermist. From 1859 until his death he was the superintendent of the London Zoo and published numerous articles on his observations of zoo animals.

Bastian, Adolf Philipp Wilhem (1826–1905): German ethnologist, explorer, and founding director of the Museum für Völkerkunde in Berlin (today the Museum of Ethnology). Bastian studied law, medicine, and natural sciences and used his inherited wealth for extended research expeditions that took him around the world several times. He played a major role in the organizational and theoretical founding of German ethnology and was chairman of the Gesellschaft für Erdkunde zu Berlin (Berlin Geographical Society), and in that capacity he also initiated the establishment of the Deutsche Gesellschaft zur Erforschung Äquitorial-Afrikas (German Society for the Exploration of Equatorial Africa).

Battel(l), Andrew (ca. 1565–1614): English pirate who spent about eighteen years in Angola as a prisoner of the Portuguese and as a trader. His adventurous travelogue was recorded by the clergyman Samuel Purchas and published in 1613. Although that text included some information about the gorilla, it was only the second, substantially expanded edition (published in 1625) that contained the description quoted in this book.

Benn, Gottfried (1886–1956): German doctor, poet, and essayist, considered one of the most important German writers of literary modernism.

Bodinus, Heinrich Karl August (1814–1884): German physician, zoologist, and zookeeper. After completing his university studies he initially worked as a doctor until 1859, when he was appointed director of the zoo in Cologne, which was in the process of being set up. In June 1869, he accepted an offer to become zoo director in Berlin, where he successfully remodeled the Zoological Garden.

Boehr, Max Ludwig August (1830–1879): German physician. Following his training and work as a military physician, Boehr became physician of the district of Nieder-Barnim. In the Berlin address book of 1877, Boehr, who was appointed Königlicher Sanitätsrat toward the end of his life, was listed as "Wundarzt und Geburthelfer" (surgeon and obstetrician).

Bolau, Heinrich (1836–1920): German zoologist and zoo director. Bolau, who studied natural sciences, was director of the Hamburg Zoological Garden from 1875 to 1909. In his study *Die menschenähnlichen Affen des Hamburger Museums* (The humanoid apes of the Hamburg Museum), written in 1876 with the anatomist Adolf Pansch (1841–1887), he dealt almost exclusively with gorillas.

Boucourt, Marie-Firmin (1819–1904): French zoologist, explorer, and draftsman. Staff member of the Muséum national d'Histoire naturelle.

Brandes, Georg Morris Cohen (1842–1927): Danish literary critic, philosopher, and writer. Brandes lived in Berlin from 1877 to 1883 and captured his impressions from that time in several reports, which were also published in book form in Germany in 1889.

Brehm, Alfred Edmund (1829–1884): German zoologist and zookeeper, writer, and explorer. Brehm's name became world-famous through his multivolume work *Brehms Tierleben* (*Brehm's Life of Animals*), published in numerous editions and synonymous with zoological literature presented as popular science. The Berlin Aquarium "Unter den Linden" was built under his leadership, and he was director of the institution from 1869 to 1873/74.

Broesike, Gustav (1853–1923): German physician and anatomist. Broesike, who studied medicine in Königsberg and Berlin, served two years in the Turkish army before becoming assistant at the Anatomical Institute of Berlin University in 1877. He dissected not only Pongo but also other apes that died at the Aquarium "Unter den Linden." He was awarded the title of professor in 1918.

Buckland, Frank (Francis Trevelyan) (1826–1880): English surgeon and journalist. Buckland, who studied at Oxford, reached a wide readership with his popular books, such as *Curiosities of Natural History*. He was also the publisher and editor of the journal *Land and Water* and was interested primarily in fish and other sea creatures.

Darwin, Charles Robert (1809–1882): English naturalist and explorer, whose theory of evolution fundamentally transformed the existing picture of the world. His 1859 book *On the Origin of Species by Means of Natural Selection, or the Preservation of Favoured Races in the Struggle for Life* is considered one of the most important works in the history of science.

Du Chaillu, Paul Belloni (1831/1835–1903): Franco-American Africa explorer. Du Chaillu was the son of a Parisian merchant who did business in West Africa. His travelogue *Explorations and Adventures in Equatorial Africa* attracted attention around the world because of its description of the gorilla. Since the veracity of his account was challenged, he undertook a second expedition in 1863. He spent the years between 1871 and 1878 in Sweden, Lapland, and northern Finland, writing several more books about his experiences there.

Düringen, Bruno (1853–1930): German poultry scientist. Dürigen was an autodidact and wrote on numerous zoological topics. Among other things, he published a two-volume manual and textbook on poultry science, and he worked for many years as an editor of scientific journals.

Falkenstein, Julius August Ferdinand (1842–1917): German physician and Africa explorer. Falkenstein, whose father was also a physician, attended the French Gymnasium in Berlin and studied medicine at the Pépinière. As a military doctor he served in the Franco-Prussian War of 1866 and the Franco-German War of 1870–71. From 1873 to 1876 he was a member of the Loango Expedition of the German Society for the Exploration of Equatorial Africa. After his return he went through the usual stages of being a Prussian military doctor, and after 1890 he ran a private practice in Berlin.

Fischer, Johann von (1850–ca. 1901): German zoologist and one of the founders of terrarium science. Fischer, about whom little is known, lived in Gotha circa 1876–77, where he evidently ran a pet shop. After 1880 he was director of the Düsseldorf Animal Park and later director of the Laboratoire d'Erpétologie in Montpellier. He published a manual of terrarium science (1884).

Fossey, Dian (1932–1985): American ethologist and environmentalist. Fossey achieved worldwide fame with her field studies of mountain gorillas and a film of her life story (*Gorillas in the Mist*). Until her murder in 1985, she advocated for the protection of animals in their natural habitats.

Frémiet, Emmanuel (1824–1910): important French sculptor of the nineteenth century. He created many animal sculptures in the style of Naturalism and Neo-Classicism, including two gorilla sculptures (1859–1887).

Freytag, Karl (1831–1908): German agronomist and expert in zoo technology. Freytag was initially the administrator of an agricultural estate, which he also used for scientific purposes. Beginning in 1875 he taught animal breeding and agrarian accounting at the University of Halle as an associate professor. In 1877, he delivered a lecture about his visit with Pongo to the Association for Natural Sciences in Saxony and Thuringia.

Gaul, August Georg (1869–1921): German sculptor. After Gaul had won a free pass to the Berlin Zoo in 1890, he did drawing studies there for years and created several sculptures of the orangutan Jumbo. His work *Running Orangutan*, created in the Neo-Baroque style, is a cast made after the artist's death from a plaster sculpture in the Berlin Museum of Natural History.

Geoffroy Saint-Hilaire, Isidore (1805–1861): French zoologist. After 1841 he was professor at the Musée national d'Histoire naturelle, and after 1850 at the University of Paris. Thanks to the gorilla specimens brought to France, Geoffroy Saint-Hilaire realized in 1852 that the gorilla was a separate species. In 1860, he founded the Jardin d'Acclimatation.

Gerlach, Christian Andreas (1811–1877): German veterinarian. Gerlach was initially professor of veterinary medicine in Hannover and, after 1870, professor in Berlin and director of the School of Animal Medicine.

Güßfeldt, Paul Richard Wilhelm (1840–1920): German geographer, explorer, and alpinist. After attending the French Gymnasium in Berlin, Güßfeldt studied mathematics and natural sciences; following his habilitation in 1868, he worked as private lecturer. He served as a volunteer in the Franco-German War. At the beginning of 1873, he was put in charge of the Loango expedition. Following his return to Germany in the summer of 1875, he undertook several expeditions, and as an outstanding alpinist he achieved numerous first ascents (in the Andes, for example). In 1892, he was appointed professor at the Oriental Seminar of the University of Berlin.

Hagenbeck, Carl Gottfried Wilhelm Heinrich (1844–1913): German wild animal dealer, organizer of ethnological shows, and zoo director. Hagenbeck, who was extremely popular, revolutionized zoo architecture by inventing panorama-like animal enclosures without bars in his animal park in Hamburg-Stellingen, which opened in 1907.

Halwas, Adolf Johann Gustav (1836–ca. 1919): German photographer. Halwas made a name for himself in the German-Danish War of 1864 as an "event photographer." In early 1877, he opened a studio at Friedrichstraße 185. He was a member of the "Photographic Society" of Berlin and carried the title of a court photographer to the duke of Saxony-Meinigen. He was considered conscientious and skillful, and he also worked as a portrait, architectural, and theater photographer.

Hanno (before 480–440 B.C.E.): also known as Hanno the Navigator; Carthaginian explorer and admiral who, circa 470 B.C.E. (according to some sources thirty to fifty years later), sailed along the western coast of Africa as far as the Gulf of Guinea to open up new trading routes. His travelogue (*Periplus*) has come down to us in a Greek translation.

Hartmann, Robert Karl Eduard (1831–1893): German physician, zoologist, Africa explorer, and ethnologist. As professor of anatomy at Berlin University, he published, among other things, a series of summarizing works about apes, one of which was the 1880 monograph *Der Gorilla: Zoologisch-Zootomische Untersuchungen* (The gorilla: Zoological-zootomical studies).

Hermes, Otto (1838–1910): pharmacist, chemist, and politician. Hermes developed a process for creating artificial seawater, which was first deployed at the Berlin Aquarium "Unter den Linden." As the successor to Alfred Brehm, Hermes ran the Aquarium from 1874 until his death, putting not only the gorilla Pongo but also other apes on display. As a liberal politician (Deutsche Fortschrittspartei / Deutsche Freisinnige Partei), he was, after 1881, a member of the German Reichstag for nearly three decades, and prior to that a member of the Prussian Landtag and city councilor of Berlin.

Hirschfeld, Paul (biographical data unknown): German writer. In 1876, Hirscheld wrote for the *Illustrirte Sonntagsblatt*, which also published his piece about Pongo. In addition, he was the author of the *Gartenlaube*. (He is not the Paul Hirschfeld who published numerous works on economics.)

Huxley, Thomas Henry (1825–1895): English biologist, explorer, and educator. Huxley was among the most influential supporters of Darwin in the struggle to establish the theory of evolution (he was referred to as "Darwin's bulldog"). His 1863 collection of essays titled *Evidence as to Man's Place in Nature* was the starting point of intense controversies on the question of descent. Huxley is considered one of the leading proponents of agnosticism, a term he coined and established.

Junhold, Jörg (1964–): German veterinarian and zoo director. Under his leadership, the Leipzig Zoo is currently implementing the concept of the "Zoo of the Future." In October 2011, Junhold was elected president of the World Association of Zoos and Aquariums (WAZA).

Kollmann, Arthur (1858–1941): German physician. Kollmann, who worked as a practicing physician, police physician, and director of a clinic for skin diseases, wrote numerous scientific works, and in 1901 he was appointed to an associate professorship for diseases of the urinary organs at the University of Leipzig. He is considered the doyen of the German study of puppetry.

Koppenfels, Hugo Kobe von (1834–1884): German gentleman farmer and Africa explorer ("gorilla hunter"). Koppenfels sold his estate near Leipzig in order to hunt and study gorillas and other large animals in western and central Africa beginning in 1873. He published highly regarded reports about his activities in the journal *Gartenlaube*, and in his lectures he showed specimens of the animals he had hunted. Hugo von Koppenfels died as the result of an injury he suffered during his third expedition to Africa.

Lauer, Gustav Adolph von (1808–1889): physician of the Prussian General Staff and personal physician to the Prussian king and German emperor Wilhelm I. Following his education and training at the Pépinière, this son of an impoverished Protestant pastor rose to the highest military offices and was ennobled in 1866.

Lenz, Oskar (1848–1925): German-Austrian Africa explorer, geologist, and mineralogist. Lenz studied in his native city of Leipzig, where he also obtained his doctorate. Following employment at the k.k. Geologische Reichsanstalt (Imperial Royal Geological Institute) in Vienna, he took Austrian citizenship. From 1874 he participated in the expedition organized by the German Society for the Exploration of Equatorial Africa and led two further Africa expeditions later on. In the summer of 1887, he was given a professorship of geography at the University of Prague.

Leutemann, Heinrich Gottlob (1824–1905): German animal painter and author. Leutemann studied at the Art Academy in Leipzig and worked in the second half of the nineteenth century for, among others, the journals *Gartenlaube* and *Illustrirte Zeitung*. In the process he also drew Pongo and the "Pinkertonian orangutan." Through his friendship with Carl Hagenbeck he was a frequent guest in Hamburg, where he occupied himself primarily with Hagenbeck's animal and ethnological shows.

Liebermann, Max (1847–1937): German painter and graphic artist, from 1920 to 1933 president of the Prussian Academy of Arts.

Lindner, Otto (1852–1945): German mechanic. Lindner was a member of the German Loango expedition from 1873 to 1876, where his chief responsibility was caring for the research station. Shortly after the end of the expedition he returned to the Congo region, where he worked with Henry Stanley for the Belgian king Leopold II.

Livingston, David (1813–1873): Scottish missionary and Africa explorer. All told, Livingston, who was living in Africa since 1841, undertook three large expeditions. In the process he was the first to cross all of Africa from east to west; he mapped the course of the Zambezi; and he named Victoria Falls. The search for this world-famous explorer, who was believed to have "disappeared" after 1869, attracted international attention.

Lloyd, William Alford (1826–1880): English bookbinder and aquarium builder. Following his apprenticeship, Lloyd devoted himself to the emerging field of aquaristics and came up with numerous technical innovations that would allow for the operation of display aquariums (e.g., water circulation and ventilation systems). Lloyd oversaw the creation of aquariums in Hamburg, Paris, Naples, London, and elsewhere. Since the Englishman insisted on a simple architectural design for the visitor spaces, Alfred Brehm rejected his already existing plans for the construction of the Berlin Aquarium.

Lüer, Wilhelm Johann Heinrich (1834–1870): German architect. Lüer studied at the Polytechnical School in Hannover, where he became a private lecturer for Architecture and Ornamentation in 1869. A substantial portion of his private commissions consisted of planning and building zoological facilities, as for example in the Hamburg Zoo, the "Egestorf Aquarium," and the Berlin Aquarium "Unter den Linden." His architectural style was characterized by Neo-Gothicism and the use of artificial rocks and grottoes. Mentally ill, Lüer took his own life in 1870.

Meyer, Adolf Bernhard (1840–1911): German scientist, explorer, and anthropologist. After 1874 he was director of the Royal Museum of Natural History in Dresden. In addition to his ornithological studies, he also took an interest in primates, including apes. In the so-called Mofaka quarrel, he was one of the chief advocates of the "chimpanzee thesis."

Meyerheim, Franz (1838–1880): German painter. Like his brother Paul Meyerheim, Franz studied at the Art Academy in Berlin. While he also painted animals, other subjects predominated in his works, which were well received by the official art critics.

Meyerheim, Paul Friedrich (1842–1915): German painter and graphic artist. Meyerheim attended the Art Academy in Berlin. He became known above all for his precise portraits of exotic animals, which he studied chiefly in zoological gardens and traveling menageries. For example, he traveled to Dresden to paint the female chimpanzee Mofaka.

Mivart, St. George Jackson (1827–1900): English zoologist and Catholic natural philosopher. As an anatomist, Mivart devoted himself to the comparative study of monkeys and changed from a supporter into an ardent opponent of Darwin's theory of evolution. His efforts to reconcile Catholic doctrine with the natural sciences initially won him wide recognition by the church, but eventually led to his excommunication.

Moore, Thomas John (1824–1892): English curator and museums director. At the age of nineteen, Moore became the assistant to the 13th Earl of Derby, a politician, naturalist, and art collector. In 1851, he became curator and later director of the Free Public Museum of Liverpool (Derby Museum), which had emerged from the earl's collection.

Mützel, Gustav Ludwig Heinrich (1839–1893): German animal painter. Mützel studied at the Art Academy in Berlin and became known above all for his illustrations for the second edition of *Brehm's Life of Animals*. He visited Pongo for the first time on 3 July 1876 in the Aquarium "Unter den Linden," and he created various woodcuts and chromolithographs of the gorilla based on photographic scenes.

Nachtigal, Gustav (1834–1885): German physician and Africa explorer. Following his study of medicine, Nachtigal initially worked as a military doctor. Beginning in 1863 he spent time in North Africa to cure a case of tuberculosis. In the following years he undertook extensive expeditions into the eastern Sahara and the eastern part of Sudan, which made him internationally known. Beginning in 1875 he was chairman of the Africa Society. In 1882, he was appointed consul general of Tunis, and in 1884, imperial commissioner for German West Africa.

Nissle (Nißle), Carl (1839–?): German philologist and historian. In the 1870s, Nissle, about whom little is known, published numerous articles about apes and the problems of zoos and aquariums in *Gartenlaube*, *Zoologischer Garten*, and other journals.

Owen, Richard (1804–1892): English zoologist, anatomist, and paleoanthropologist. Next to Charles Darwin, Owen was considered the most important naturalist of the Victorian era. Owen promoted the establishment of a separate Museum of Natural History, whose first director he was; wrote many important works on the comparative anatomy of vertebrates (including apes); and coined the term "Dinosauria" in 1841.

Pechüel-Loesche, Eduard Moritz (1840–1913): German geographer, zoologist, and Africa explorer. Following the early death of his parents, Pechüel-Loesche traveled the world as a seaman in the merchant marine before studying sciences and philosophy and earning his doctorate with a zoological dissertation. As part of his participation in the German Loango expedition, he took over the meteorological station in Chinchoxo. Between 1882 and 1883 he acted as Henry Stanley's agent in the Congo. In 1895, he assumed a professorship for geography at the University of Erlangen.

Peters, Wilhelm Carl Hartwig (1815–1883): German zoologist and Africa explorer. The son of a pastor, he studied medicine and natural history and undertook a long study trip to Africa between 1842 and 1847. After 1857, Peters, who wrote nearly four hundred articles about the most varied animal species, ran the Museum of Zoology of Berlin University as a university professor.

Pinkert, Ernst Wilhelm (1844–1909): German innkeeper and founder as well as long-time director of the Leipzig Zoo. Between 1893 and 1895, Pinkert displayed a total of six so-called giant orangutans, four of whom (Max, Moritz, Anton, and Jumbo) attracted particular attention because of their size and cheek pads.

Rabl-Rückhard, Hermann (1839–1905): German physician and anatomist. Following his training as a military doctor, Rabl-Rückhard was given a position as assistant at the Anatomical Institute of Berlin University in 1875 and was appointed professor of anatomy in 1884.

Reichert, Bogislaus Karl (1811–1883): German physiologist and anatomist. Reichert studied medicine in Königsberg and Berlin and held professorships at various universities. Beginning in 1858, he was a professor at Berlin University. He is considered one of the founders of modern developmental biology. In addition, he studied the anatomy of the brain, the development of the skull (including that of apes), and histology.

Savage, Thomas Staughton (1804–1880): American missionary, doctor, and naturalist. In 1836, Savage was sent to Liberia, where he worked as a missionary and doctor and as a naturalist on the side. As early as 1844 he published an article about chimpanzees together with Jeffries Wyman. Thanks to his article about gorillas published in December 1847 (also in collaboration with Wyman), Savage is considered the scientific discoverer of these apes.

Schaller, George Beals (1933–): American zoologist and environmentalist. Schaller was the first scientist to study mountain gorillas in the wild and published several books, including *The Year of the Gorilla* (1966).

Schmidt, Max (Maximilian) (1834–1888): German veterinarian and zoo director. Following his studies, Schmidt worked initially as a veterinarian. He was appointed managing director of the Frankfurt Zoo in 1859 and its director in 1864. He contributed to the development of the foundations of scientific animal keeping through a great many publications. They included an article about apes and an 1870 book about the diseases of apes. Beginning in 1885, Schmidt was director of the Berlin Zoo.

Schoepf, Alwin (1823–1881): German apothecary and, from 1861 until his death, inspector and then director of the Dresden Zoo.

Schweinfurth, Georg (1836–1925): German botanist and Africa explorer. Schweinfurth studied botany and paleontology, and beginning in 1861 he undertook extensive research trips to Africa, which brought him to, among other places, the region of the Upper Nile and the Congo. Schweinfurth also advocated the establishment of a German colonial empire.

Soyaux, Herman (1852–after 1928): German botanist, Africa explorer, and head of a settlement colony in Brazil. Following his study of botany, Soyaux became a member of the German Loango expedition and arrived at the German research station of Chinchoxo in January 1874. In 1875, he undertook a journey into the interior of Angola. After his return to Europe, he published several books. He went to Gabon in 1879 for a longer period of time, and in 1888 to Brazil, where he seems to have spent the rest of his life.

Stanley, Henry Morton (1841–1904): British-American journalist and Africa explorer. He became famous for his quest to find David Livingston and through the ruthless colonization of the Congo.

Thomas, Friedrich (?–ca. 1899): German sculptor. Thomas ran a studio at Oranienstraße 124 in Berlin and built a villa in Waren (Müritz) in 1884, where he became acquainted with, among others, the family of Theodor Fontane.

Viereck, Wilhelm (1849–1924): German animal keeper. According to the Berlin address book, he worked for the Aquarium "Unter den Linden" from 1875 to 1879 and, with his wife, was responsible for Pongo's care.

Virchow, Rudolf Ludwig Karl (1821–1902): German physician, archaeologist, politician, and founder of modern pathology. Virchow, who taught chiefly at Berlin University, is considered the champion of a strictly scientific and socially oriented medicine and one of the most important modern physicians. His numerous fields of work included anthropology and ethnology, and one topic that interested him was the shape of the skull in humans and apes.

Ward, Roland (1835–1912): English taxidermist. Ward, who had learned the craft from his father, opened his own workshop in London in 1872 and was among the leading taxidermists before the First World War. He arranged numerous exhibitions and wrote several books, including one about the hunting of wild animals.

Waterton, Charles (1782–1865): English naturalist, traveler, and eccentric. Waterton, who continuously attracted attention with unconventional ideas and behaviors, initially made a name for himself through travel accounts (British Guyana and Brazil). Later, on his country estate in Walton Hall, he established what was probably the first nature preserve in the world. The observations about the gorilla Jenny appeared in 1857 in the third volume of his *Essays on Natural History*.

Wickersheimer, Jean Georg (1832–1896): German taxidermist. Wickersheimer, who hailed from Strasbourg, initially completed an apprenticeship as a carpenter in Paris and later became an employee at the Anatomical Institute of Berlin University. There he rose to become a taxidermist in 1878/79 and made a name for himself with a patent "for the dry conservation and the wet storage of organic objects."

Wolf, Joseph (1820–1899): Anglo-German animal painter. At the request of Richard Owen, Wolf, who is frequently described as the most eminent animal painter of the nineteenth century, also drew the portrait of a large gorilla cadaver that had been brought to London in a barrel of alcohol in 1858.

Wyman, Jeffries (1814–1874): American scientist and anatomist. Wyman studied in Europe for several years (with Richard Owen, among others) and was professor of anatomy at Harvard College from 1847 until his death.

NOTES

CHAPTER 2

1. *The Strange Adventures of Andrew Battell of Leigh, in Angola and the Adjoining Regions: Reprinted from "Purchas His Pilgrimes"* (London, 1901), 54–55.

2. See *Boston Journal of Natural History* 5, no. 4 (1847): 417–43. Thomas Savage's discovery had been communicated in August 1847 in the same journal with a brief notice.

3. While nineteenth-century scholars often asserted that the term "gorilla" was a mix-up with the name "Gorgons," today we know that it derives from the language of the Hausa-Fulani and means something like "little man" or "pygmy."

4. The fact that Frémiet depicted a female animal did nothing to prevent the scandal. The sculpture is attested only in two photographs, which are today in the Musée d'Orsay in Paris.

5. On this, see the very informative book by Julia Voss, *Darwins Bilder: Ansichten der Evolutionstheorie, 1837–1874* (Frankfurt am Main, 2007), 240–41, 291ff.

6. Quotation from the original English edition of Paul Du Chaillu, *Explorations and Adventures in Equatorial Africa* (New York, 1868), 98–99; originally published London, 1861; translated into German as *Reisen in Zentralafrika* (Berlin, 1862).

7. Charles Waterton, *Essays on Natural History, Third Series* (London, 1857), 65–66. Available on the internet at https://archive.org/stream/essaysonnaturoowate#page/64/mode/2up.

8. In 2009, the zoologist and filmmaker Karen Partridge created a radio feature for the BBC in which she investigated the story of Jenny.

CHAPTER 3

1. The image of the gorilla is dated 12 October 1875. In addition to the album *Die Loango-Küste in 72 Original-Photographien (35 Blatt) nebst erläuterndem Texte von Dr. Falkenstein* (Berlin, 1876), another album in three installments was published that same year, reserved for members of the German Society for the Exploration of Equatorial Africa. It, too, contains two photographs of Pongo.

2. The business premises of the Society, whose membership was made up largely of scientists, merchants, civil servants, and physicians, were located in the apartment rented by the Berlin Geographical Society for its library at Kronenstraße 21, and from April 1874 at Krausenstraße 42.

3. Only later did the Kaiser raise the sum paid from his disposition fund from one thousand to twenty-five thousand talers. Paid several times and covering most of the costs, these funds did not, however, allow for any long-term calculation, since they did not represent any legal and binding commitment.

4. Having returned to Germany, Bastian published a two-volume account of the expedition in 1875, in which he took up the gorilla issue one more time and wrote this: "Mr. Alcantara kept a young gorilla (Pongo) at the Chicambo station for four months. This gorilla (as he told it) was completely trusting and settled in after a short time, bringing fire from the kitchen, standing upright at the table, covering his head with a cap, sleeping on a bed of pads at night, and if it was not there, demanding it with extended screams." Adolf Bastian, *Die deutsche Expedition an der Loango-Küste*, vol. 1 (Jena, 1874), 245.

5. The best overview of the history of the expedition is by Beatrix Heintze; see the bibliography.

CHAPTER 4

1. See the *Correspondenzblatt der Afrikanischen Gesellschaft*, no. 15 (1875): 256.

2. In the language of the Bavili (in the Bantu language group), *Mpungu* describes the person who is the ruler or holds absolute power. A *Mpungu*, that is, a gorilla, is thus a king of the forest.

3. Excerpts of the letter were published in the *Correspondenzblatt der Afrikanischen Gesellschaft* (Berlin, 1876): 304–7, and also in the *Zeitschrift für Ethnologie* (Berlin, 1876): 60–61.

4. There had been earlier photographs of living gorillas, though their whereabouts are unknown: there was the photo of Jenny from 1855/56, that

of an English trader in Gabon from the period shortly after 1861, and that of Paul Du Chaillu from 1876.

CHAPTER 5

1. Although the presumably very brief letter from Charles Darwin is not listed in the Darwin Correspondence Project, there is no reason to doubt Julius Falkenstein's statement. Falkenstein was too upstanding a person to invent this story, which he repeated several times.

2. These numbers refer to the period around 1876. Before that, young chimpanzees had cost more; later, their price dropped, at times sharply.

3. Excerpts from Nachtigal's letter were available on the internet for a while, before Kotte Autographs GmbH sold it to an unnamed client. Unfortunately, my attempts to contact the buyer were unsuccessful.

CHAPTER 6

1. The published information about Otto Hermes is contradictory when it comes to specific details. His biography was most accurately assembled by Harro Strehlow in 1987. The archive of the Leopoldina also has a short questionnaire that Hermes filled out by hand when he joined the Academy in 1893. In it we are told, among other things, that his father was the landowner Carl Hermes in Meyenburg, that he attended the Real Gymnasium in Perleberg, studied in Jena and Berlin from 1858 to 1862, and obtained his doctorate in Leipzig in 1866. The chemical factory that Hermes owned from 1864 to 1871 was located at Ritterstraße 35. As a liberal member of parliament, he fought, for example, against the discrimination against Jews in the military, for the development of deep-sea fishing, and for the protection of birds.

2. On this, see the numerous articles by Harro Strehlow, who has devoted a lot of time to the history of the Aquarium "Unter den Linden." At the time the Aquarium was called only the Berlin Aquarium, but today it is generally referred to as the Aquarium "Unter den Linden."

3. In a few rare cases, traveling menageries also had chimpanzees or orangutans. It is not known whether they were exhibited in Berlin.

CHAPTER 7

1. Alfred Brehm stated that the presence of eight hundred visitors at the same time already created "a real crush" around the aviaries and the Aquarium's show tank.

2. Quoted from Karl August Specht, *Theologie und Wissenschaft oder alte und neue Weltanschauung*, 3rd rev. and exp. ed. (Gotha, 1878), 257.

3. Compared to 1875, which was already considered a good year, attendance was up by about 70,000. One should bear in mind that Berlin's population was still around 980,000 at that time, and that the admission fees to the Aquarium were fairly high and were rarely reduced.

4. Otto Hermes, "Der Gorilla und seine nächsten Verwandten," *Der Zoologische Garten*, no. 1 (1877): 58–59.

5. It is difficult to judge whether the logs that are sometimes depicted were part of the decoration. None of the reports about the gorilla mention them.

6. Prints of the photographs are in the papers of Eduard Pechuël-Loesche at the Leibniz-Institut für Länderkunde. The photographer Heinrich Lichterfeld also took a picture of Pongo.

7. We learn from the yearbooks of the Nassauischer Verein für Naturkunde that Wiesbaden purchased the plaster busts of Molly, the orangutan, and Pongo.

CHAPTER 8

1. Otto Hermes, "Der Gorilla und seine nächsten Verwandten," *Der Zoologische Garten*, no. 1 (1877): 58–59.

2. See the bibliography.

3. Johannes von Fischer, "Besuch bei M'Pungu," *Der Zoologische Garten*, no. 3 (1877): 167.

4. The health council included the professors Rudolf Virchow, Christian Gerlach, Karl Bogislaus Reichert, and Wilhelm Peters and the physicians Paul Langerhans and Julius Falkenstein. Since Robert Hartmann, Otto Hermes, and the physician Martini are also mentioned in connection with the gorilla's illnesses, the circle of experts had at least nine members.

5. We read the following about the chimpanzee Tschego, who died in 1876 at the Aquarium "Unter den Linden": "However, for a longer period before its death, those around him noticed a terrible stench, which undoubtedly emanated from its mouth. They then saw that the gums were starting to rot and soon it also lost a few incisors, of which eventually only a few remained in place, while the area around the molars was slowly affected by the same disease from the front to the back. Over time the appetite was lost, and there was diarrhea now and then. Usually a picture of the most cheerful life, the animal soon grew surly, though very resigned and surrendering to its sad fate. From time to time there were severe fits of coughing, which seemed

to increase significantly in number during the last few days of its life, and the animal died on 10 September with the most severe signs of suffocation." Summary of meeting on 21 November 1876 in *Sitzungsberichte der Gesellschaft Naturforschender Freunde zu Berlin* (1876): 140.

6. H. Schneider, "Ueber Erhaltung der anthropomorphen Affen," *Der Zoologische Garten* 22, no. 2 (1881): 47.

CHAPTER 9

1. (Gustav) Lunze, "Der Gorilla des Berliner Aquariums und seine Reise nach London," Der Garten 19, no. 3 (1878): 90–91. Gustav Lunze, who summarized the account by Hermes, was a member of the Berliner Gesellschaft für Anthropologie, Ethnologie und Urgeschichte, and in 1877 he published a book entitled *Die Hundezucht im Lichte der Darwinschen Theorie* [Dog breeding in light of the Darwinian theory].

2. I analyzed about a dozen papers and magazines, including *the Examiner*, the *Graphic*, and *Lloyd's Weekly Newspaper*. The *Illustrated London News* and the *Illustrated Police News* also published drawings of Pongo, for example, one by the very well-known illustrator and painter John Gilbert (1817–1897). Another image of Pongo appeared in 1886 in John Fortuné Nott, *Wild Animals Photographed and Described* (London, 1886), 536.

3. Frank Buckland, *Notes and Jottings from Animal Life* (London, 1882), 10–11.

4. *Punch*, 4 August 1877, 41. The caricatures were from the pen of Edward Linley Sambourne (1844–1910).

CHAPTER 10

1. See Ernst Maire, *Der Gorilla* (Berlin, 1877). I was unable to determine whether the name Ernst Maire, which is not otherwise attested, is a pseudonym. The description itself follows in large sections the published accounts of Otto Hermes and cost thirty pfennigs.

2. *Berliner Klinische Wochenschrift* 14, no. 48 (26 November 1877).

3. Alfred Edmund Brehm, *Brehms Thierleben: Allgemeine Kunde des Thierreichs; Große Ausgabe in 10 Bänden. Zweite umgearbeitete und vermehrte Auflage* (Leipzig, 1876), 1:56–57.

BIBLIOGRAPHY

ARCHIVAL SOURCES

Archiv der Leopoldina—Nationale Akademie der Wissenschaften
MM 3019 Biographischer Fragebogen Otto Hermes
Bayerische Staatsbibliothek, Nachlässe und Autographen
Die Tagebücher von Eduard Pechuël-Loesche von seiner Reise an die Loangoküste (24.2.1875–18.6.1876)
Kleines Schloss Blankenburg (Harz)
Zeichnungen und Fotografien aus dem Nachlass von Robert Hartmann
Leibniz-Institut für Länderkunde, Archiv für Geographie
Nachlass Eduard Pechuël-Loesche, 227/32
Museum für Naturkunde Berlin
Nr. 85777, gorilla, skull, skeleton
Staatsbilbiothek zu Berlin, Nachlässe und Autographen
Slg. Darmst. Afrika 1873: Falkenstein, Julius

PERIODICALS

Annals and Magazine of Natural History (London, 1861–62)
Archives du Muséum d'Histoire Naturelle (Paris, 1858–61)
Archiv für Naturgeschichte (Berlin, 1852–61)
Aus der Heimat (Leipzig, 1865)
Berliner Adressbuch (Berlin, 1873–79)
Berliner Fremdenblatt (Berlin, 1876–77)
Berliner-Gerichtszeitung (Berlin, 1876–77)
Berliner Klinische Wochenschrift (Berlin, 1877)
Berliner Tageblatt (Berlin, 1876–77)
Boston Journal of Natural History (Boston, 1847)
Bulletin de la Société Linnéenne de Normandie (Caen, 1862)
Bulletin du Muséum nationale d'histoire (Paris, 1920)
Correspondenzblatt der Afrikanischen Gesellschaft (Berlin, 1873–76)
Daily News (London, 1877)
Deutsche Bauzeitung (Berlin, 1869)

Deutsche Roman-Zeitung (Berlin, 1876)
Dresdner Nachrichten (Dresden, 1875)
L'Explorateur: Journal Géographique et Commercial (Paris, 1875–76)
Die Gartenlaube (Leipzig, 1865–81)
Gorilla-Journal (Mühlheim and Ruhr, 1999–2011)
Graphic (London, 1876)
Hamburgischer Correspondent (Hamburg, 1876)
Hardwick's Science Gossip (London, 1876–77)
Illustrated London News (London, 1877)
Illustrirte Zeitung (Leipzig, 1853–80)
Isis: Zeitschrift für alle naturwissenschaftlichen Liebhabereien (Berlin, 1876–77)
Jahrbuch des Nassauischen Vereins für Naturkunde (Wiesbaden, 1878–79)
Journal Amusant (Paris, 1859)
Journal des Voyages et des Aventures de Terre et de Mer (Paris, 1877–78)
Journal für Gasbeleuchtung und verwandte Beleuchtungsarten sowie für Wasserversorgung (Munich, 1873)
Leipziger Nachrichten (Leipzig, 1877)
Leipziger Tageblatt und Anzeiger (Leipzig, 1876)
Leopoldina: Amtliches Organ der Kaiserlich Leopoldinisch-Carolinischen Deutschen Akademie der Naturforscher (Dresden, 1875)
Liverpool Mercury (Liverpool, 1876–77)
Mitteilungen der Gesellschaft für Erdkunde zu Leipzig (Leipzig, 1877)
Mittheilungen aus Justus Perthes' Geographischer Anstalt über wichtige neue Forschungen aus dem Gesammtgebiete der Geographie von Dr. A. Petermann (Gotha, 1875–77)
Die Natur, new series (Halle, 1876–77)
Nature (London, 1875–77)
Naturwissenschaftliche Wochenschrift (Berlin, 1887–95)
New York Times (New York, 1875–77)
19th Century British Library Newspapers (London, 1876–77)
Popular Science Monthly (New York, 1877–78)
Proceedings of the Academy of Natural Science of Philadelphia (Philadelphia, 1878)
Proceedings of the Zoological Society of London (London, 1848–50, 1852–53, 1860, 1862, 1863, 1875–78)
Punch (London, 1861, 1877)
Revue d'Anthropologie (Paris, 1878)
Sitzungsberichte der Gesellschaft Naturforschender Freunde zu Berlin (Berlin, 1876–77)

Sitzungsberichte der Königlich Preussischen Akademie der Wissenschaften zu Berlin (Berlin, 1882)
Times (London, 1876, 1877)
Transactions of the Zoological Society of London (London, 1849, 1866)
Die Tribüne (Berlin, 1876)
Ueber Land und Meer: Allgemeine Illustrirte Zeitung (Stuttgart, 1876)
Velhagen & Klasings Monatshefte (Leipzig, 1921–22)
Verhandlungen der Berliner Gesellschaft für Anthropologie, Ethnologie und Urgeschichte (Berlin, 1872–76)
Verhandlungen der Gesellschaft für Erdkunde zu Berlin (Berlin, 1874–77)
Vossische Zeitung (*Königlich privilegierte Berlinische Zeitung von Staats- und gelehrten Sachen*) (Berlin, 1876–77)
Westermann's Jahrbuch der Illustrierten Deutschen Monatshefte (Brunswick, 1870, 1876, 1877)
Zeitschrift für Bauwesen (Berlin, 1869)
Zeitschrift für Ethnologie (Berlin, 1872–77, 1899)
Zeitschrift der Gesellschaft für Erdkunde zu Berlin (Berlin, 1870, 1873, 1875, 1878)
Der Zoologische Garten: Zeitschrift für Beobachtung, Pflege und Zucht der Thiere (Frankfurt am Main, 1865–92)

BOOKS AND ARTICLES

Album der Deutschen Gesellschaft zur Erforschung Aequatorial-Afrikas. Landschaftlicher Theil, I. bis III. Lieferung. Berlin, 1876.
Artinger, Kai. *Von der Tierbude zum Turm der Blauen Pferde: Die künstlerische Wahrnehmung der wilden Tiere im Zeitalter der zoologischen Gärten.* Berlin, 1995.
Bastian, Adolf. *Die deutsche Expedition an der Loango-Küste, nebst älteren Nachrichten über die zu erforschenden Länder: Nach persönlichen Erlebnissen.* 2 vols. Jena, 1874–75. Reprint, Hildesheim, 2007.
Baumuk, Bodo-Michael, and Jürgen Rieß, eds. *Darwin und Darwinismus: Eine Ausstellung zur Kultur- und Naturgeschichte.* Berlin, 1994.
Becker, Kurt. "Abriß einer Geschichte der Gesellschaft Naturforschender Freunde zu Berlin." *Sitzungsberichte der Gesellschaft Naturforschender Freunde zu Berlin,* n.s., 13, no. 1 (1973): 1–58.
Berlin und seine Bauten. Edited by the Architekten-Verein zu Berlin. 2 parts. Berlin, 1877.
Berlins Aufstieg zur Weltstadt: Ein Gedenkbuch hrsg. vom Verein Berliner Kaufleute und Industrieller aus Anlass seines 50jährigen Bestehens. Berlin, 1929.

Beta, H., ed. *Die Bewirthschaftung des Wassers und die Ernten daraus: Mit einem Vorworte von Dr. Brehm, dem Verfasser des "Illustrirten Thierlebens," des "Lebens der Vögel" u.s.w. und wissenschaftlichem Direktor des Berliner Aquariums*. Leipzig and Heidelberg, 1868.

Biella, Edgar. "Das Berliner Aquarium Unter den Linden / Schadowstraße: Zur Konzeption des Rundganges und Bedeutung des Grottenstils als Ausstellungsarchitektur." In *Der Bär von Berlin, Jahrbuch für die Geschichte Berlins*, 49th ed., edited by Sibylle Einholz and Jürgen Wetzel. Berlin, 2000.

———. "Grottenstil als Ausstellungsarchitektur des 19. Jahrhunderts: Eine Untersuchung unter besonderer Berücksichtigung des Berliner Aquariums Unter den Linden / Schadowstraße." Diplomarbeit, Berlin, 1999.

Biographisches Lexikon der hervorragenden Ärzte der letzten fünfzig Jahre. Edited by I. Fischer. Vol. 1. Berlin, 1932.

Biographisches Lexikon hervorragender Ärzte des neunzehnten Jahrhunderts: Mit einer historischen Einleitung. Edited by J. Pagel. Berlin and Vienna, 1901.

Bolau, Heinrich. *Die menschenähnlichen Affen des Hamburger Museums*. Hamburg, 1876.

Brandes, Georg. *Berlin als deutsche Reichshauptstadt: Erinnerungen aus den Jahren 1877–1883*. Translated by Peter Urban-Halle, edited by Erik M. Christensen and Hans-Dietrich Loock. Berlin, 1989.

Bredekamp, Horst, Jochen Brüning, and Cornelia Weber, eds. *Theater der Natur und Kunst: Theatrum naturae et artis*. 2 vols. Berlin, 2000. Exhibition catalog.

Brehm, Alfred Edmund. *Brehms Thierleben*. Vol. 1, *Die Säugetiere*. Große Ausgabe letzter Hand von 1876–1879. Berlin and Vienna, 1980.

———. *Brehms Thierleben: Allgemeine Kunde des Tierreichs*. Gänzlich neubearbeitet von Dr. Pechuël-Loesche. Vol. 1. Leipzig, 1900.

Brunner, Bernd. *Wie das Meer nach Hause kam: Die Erfindung des Aquariums*. Berlin, 2003.

Buchner-Fuhs, Jutta. "Das Tier als Freund. Überlegungen zur Gefühlsgeschichte im 19. Jahrhundert." In *Tiere und Menschen: Geschichte und Aktualität eines prekären Verhältnisses*, edited by P. Münch. Paderborn, 1998.

Buckland, Frank. *Notes and Jottings from Animal Life*. London, 1882.

Burton, Richard F. *Two Trips to Gorilla Land and the Cataracts of the Congo*. 2 vols. London, 1876.

Charles Darwin und seine Wirkung. Edited by Eve-Marie Engels. Frankfurt am Main, 2009.

Ciz, Karl Heinz. *Robert Hartmann (1831–1893): Mitbegründer der deutschen Ethnologie*. Gelsenkirchen, 1984.

Constant, N'Dala. "Kenosis bei Hans Urs von Balthasar als Instrument für den interreligiösen Dialog." Diss. A. Vienna, 2009.

Daum, Andreas W. *Wissenschaftspopularisierung im 19. Jahrhundert. Bürgerliche Kultur, naturwissenschaftliche Bildung und die deutsche Öffentlichkeit, 1848–1914*. 2nd supplemented ed. Munich, 2002.

Dembeck, Hermann. *Gelehrige Tiere*. Düsseldorf, 1966.

Deutsche Apotheker-Biographie. Edited by Wolfgang-Hagen Hein and Holm-Dietmar Schwarz. Vol. 1. Stuttgart, 1975.

Dittrich, Lothar and Annelore Rieke-Müller. *Carl Hagenbeck (1844–1913). Tierhandel und Schaustellungen im Deutschen Kaiserreich*. Frankfurt am Main, 1998.

Dixson, Alan F. *The Natural History of the Gorilla*. London, 1981.

Dokumentation "Berliner Fotografenateliers des 19. Jahrhunderts" FHTW—Berlin, SG Museumskunde. Berlin, n.d.

Du Chaillu, Paul. *Reisen in Zentralafrika*. Berlin, 1862. Reprinted in *150 [Einhundertfünfzig] Jahre C. Woermann: Wagnis Westafrika; Die Geschichte eines Hamburger Handelshauses, 1837–1987*. Hamburg, 1987. Orig. pub. 1861 (in English).

Erman, Adolf. *Mein Werden und mein Wirken: Erinnerungen eines alten Berliner Gelehrten*. Saarbrücken, 2007.

Essner, Cornelia. *Deutsche Afrikareisende im neunzehnten Jahrhundert: Zur Sozialgeschichte des Reisens*. Stuttgart, 1985.

Falkenstein, Julius. *Afrikas Westküste: Vom Ogowe bis zum Damara-Land*. Leipzig and Prague, 1885.

———. "De typho abdominali." Diss., Berlin, 1867.

———. "Die Loango-Expedition: Zweite Abtheilung." In *Die Loango-Expedition: Ausgesandt von der Deutschen Gesellschaft zur Erforschung Aequatorial-Africas, 1873–1876*, by Paul Güssfeldt, Julius Falkenstein, and Eduard Pechuël-Loesche. Leipzig, 1879–82.

———. *Die Loango-Küste in 72 Original-Photographien (35 Blatt) nebst erläuterndem Texte von Dr. Falkenstein*. Berlin, 1876.

Festschrift zum Hundertjährigen Bestehen der Berliner Gesellschaft für Anthropologie, Ethnologie und Urgeschichte, 1869–1969. Pt. 1, Fachhistorische Beiträge. Berlin, 1969.

Fiedermutz-Laun, Annemarie. "Adolf Bastian (1826–1905)." In *Klassiker der Kulturanthropologie: Von Montaigne bis Margaret Mead*, edited by Wolfgang Marschall. Munich, 1990.

Fischer, Manuela, Peter Bolz, and Susan Kamel, eds. *Adolf Bastian and His Universal Archive of Humanity: The Origins of German Anthropology*. Hildesheim, Zürich, and New York, 2007.

Fritzsche, Peter. *Als Berlin zur Weltstadt wurde*. Berlin, 2008.

Gautier, Jean-Pierre. "À la recherche des Gorillas." In *Coeur d'Afrique: Gorilles, cannibales et Pygmées dans le Gabon de Paul Du Chaillu*, edited by Jean-Marie Hombert and Louis Perrois. Paris, 2007.

Glatzer, Ruth. *Berlin wird Kaiserstadt: Panorama einer Metropole, 1871–1890*. Berlin, 1993.

Gondermann, Thomas. *Evolution und Rasse: Theoretischer und institutioneller Wandel in der viktorianischen Anthropologie*. Bielefeld, 2007.

Goschler, Constantin. *Rudolf Virchow. Mediziner—Anthropologe—Politiker*. Cologne, Weimar, and Vienna, 2009.

———, ed. *Wissenschaft und Öffentlichkeit in Berlin, 1870–1930*. Stuttgart, 2000.

Griem, Julika. *Monkey Business: Affen als Figuren anthropologischer und ästhetischer Reflexion, 1800–2000*. Berlin, 2010.

Güßfeldt, Paul. "Die Loango-Expedition, Erste Abtheilung." In *Die Loango-Expedition. Ausgesandt von der Deutschen Gesellschaft zur Erforschung Aequatorial-Africas, 1873–1876*, by Paul Güssfeldt, Julius Falkenstein, and Eduard Pechuël-Loesche. Leipzig, 1879–82.

———. *Mein Kriegserlebnisse im deutsch-französischen Feldzug*. Berlin, 1907.

Haemmerlein, Hans-Dietrich. *Der Sohn des Vogelpastors: Szenen, Bilder, Dokumente aus dem Leben von Alfred Edmund Brehm*. Berlin, 1985.

Haikal, Mustafa. *Die Löwenfabrik: Lebensläufe und Legenden*. Leipzig, 2006.

———. "Ein selbstgemachter Mann bester Art: Der Zoogründer Ernst Pinkert." In *Illustrierter Führer durch den zoologischen Garten zu Leipzig*, by Georg Westermann. Leipzig, 2009. Reprint of the 1883 ed.

Haikal, Mustafa, and Winfried Gensch. *Der Gesang des Orang-Utans: Die Geschichte des Dresdner Zoo*. Dresden, 2011.

Haikal, Mustafa, and Jörg Junhold. *Auf der Spur des Löwen: 125 Jahre Zoo Leipzig*. Leipzig, 2003.

Handbuch der Architektur. Vierter Teil, 6. Halbband, *Gebäude für Erziehung, Wissenschaft und Kunst*. Stuttgart, 1906.

Hartmann, Robert. "Beiträge zur zoologischen und zootomischen Kenntniss der sogenannten anthropomorphen Affen." *Archiv für Anatomie, Physiologie und wissenschaftliche Medizin* (1872).

———. *Der Gorilla: Zoologisch-zootomische Untersuchungen*. Leipzig, 1880.

———. *Die menschenähnlichen Affen*. Berlin, 1876.

———. *Die menschenähnlichen Affen und ihre Organisation im Vergleich zur menschlichen*. Leipzig, 1883.

Hauffe, Friedericke, and Heinz Georg Klös. "Der Tierillustrator Gustav Mützel." *Bongo* 26 (1995): 29–46.

Heck, Ludwig. "Menschenaffen." *Velhagen & Klasings Monatshefte* 36 (February 1922): 641–56.

Heintze, Beatrix. *Ethnographische Aneignungen: Deutsche Forschungsreisende in Angola; Kurzbiographien mit Selbstzeugnissen und Textbeispielen.* Frankfurt am Main, 1999.

———. "Feldforschungsstress im 19. Jahrhundert: Die deutsche Loango-Expedition, 1873–1876." In *Die offenen Grenzen der Ethnologie: Schlaglichter auf ein sich wandelndes Fach. Klaus E. Müller zum 65. Geburtstag,* edited by Sylvia M. Schomburg-Scherff und Beatrix Heintze. Frankfurt am Main, 2000.

Hochadel, Oliver. "Unter Menschen: Die Schimpansin Mafuka im Dresdner Zoologischen Garten (1873–75)." In *"Ich Tarzan": Affenmenschen und Menschenaffen zwischen Science und Fiction,* edited by Gesine Krüger, Ruth Mayer, and Marianne Sommer. Bielefeld, 2008.

Hübner, Rolf. *Gustav Nachtigal: Der größte Afrikaforscher 1875 in Bad Ems zum Vortrag beim Kaiser.* Bad Emser Hefte 165. Bad Ems, 1997.

Huxley, Thomas Henry. *Evidence as to Man's Place in Nature.* London, 1863.

Informationen und Berichte: Braunschweigisches Landesmuseum, no. 1 (1997). [Issue on Robert Hartmann.]

Ingensiep, Hans-Werner. "Kultur- und Zoogeschichte des Gorillas: Beobachtungen zur Humanisierung von Menschenaffen." In *Die Kulturgeschichte des Zoos,* edited by Lothar Dittrich, Dietrich von Engelhardt, and Annelore Rieke-Müller. Berlin, 2001.

———. "Mensch und Menschenaffe." In *Tiere und Menschen: Geschichte und Aktualität eines prekären Verhältnisses,* edited by P. Münch. Paderborn, 1998.

———. "Der Orang-Outang des Herrn Vosmaer: Ein aufgeklärter Menschenaffe." In *Ich, das Tier: Tiere als Persönlichkeiten in der Kulturgeschichte,* edited by Jessica Ullrich, Friedrich Weltzien, and Heike Fuhlbrügge. Berlin, 2009.

Ingensiep, Hans-Werner, and Heike Baranzke. *Grundwissen Philosophie: Das Tier.* Stuttgart, 2008.

Joseph Wolf. *Tiermaler—Animal Painter.* Edited by Karl Schulze-Hagen and Armin Geus. Marburg an der Lahn, 2000.

Junker, Thomas, and Marsha Richmond. *Charles Darwins Briefwechsel mit deutschen Naturforschern: Ein Kalendarium mit Inhaltsangaben, biographischem Register und Bibliographie.* Marburg an der Lahn, 1996.

Kaselow, Gerhild. *Die Schaulust am exotischen Tier: Studien zur Darstellung des zoologischen Gartens in der Malerei des 19. und 20. Jahrhunderts*. Hildesheim, Zurich, and New York, 1999.

Kim, Young-Ok. "Bogislaus Reichert (1811–1883), sein Leben und seine Forschungen zur Anatomie und Entwicklungsgeschichte." Diss., A. Mainz, 2000.

Klös, Heinz-Georg. "Berliner Aquariengeschichte." In *Vom Seepferdchen bis zum Krokodil*, edited by Heinz-Georg Klös and Jürgen Lange. Berliner Forum 4/85. Berlin, 1985.

———. *Von der Menagerie zum Tierparadies: 125 Jahre Zoo Berlin*. Berlin, 1969.

Klös, Heinz-Georg, Hans Frädrich, and Ursula Klös. *Die Arche Noah an der Spree: 150 Jahre Zoologischer Garten Berlin; Eine tiergärtnerische Kulturgeschichte von 1844 bis 1994*. Berlin, 1994.

Knauer, Friedrich. *Menschenaffen ihr Frei- und Gefangenleben*. Leipzig, ca. 1915.

Koch, Walter. *Bericht über das Ergebnis der Obduktion des Gorilla Bobby des Zoologischen Gartens zu Berlin: Ein Beitrag zur vergleichenden Konstitutionspathologie*. Jena, 1937.

Kohlmaier, Georg, and Barna von Sartory. *Das Glashaus: Ein Bautypus des 19. Jahrhunderts*. Munich, 1981.

Kollmann, Arthur. *Der Tastapparat der Hand der menschlichen Rassen und der Affen: Seiner Entwicklung und Gliederung*. Hamburg and Leipzig, 1883.

Koner, W[ilhelm David]. "Zur Erinnerung an das fünfzigjährige Bestehen der Gesellschaft für Erdkunde zu Berlin." *Zeitschrift der Gesellschaft für Erdkunde zu Berlin* 30 (1878): 169–250.

Krone, Dagmar. "Geographische Forschung und Kolonialpolitik: Das Beispiel der Afrikanischen Gesellschaft in Deutschland (1878–1889)." Diss. A. Magdeburg, 1984.

Krüger, Gesine, Ruth Mayer, and Marianne Sommer, eds. *"Ich Tarzan": Affenmenschen und Menschenaffen zwischen Science und Fiction*. Bielefeld, 2008.

Krüger-Kopiske, Karsten Kunibert. *Die Schiffe der Deutschen Afrika-Linien: Zeichnungen und Lebensläufe*. Hamburg, 2009.

Lange, Britta. "Die Allianz von Naturwissenschaft, Kunst und Kommerz in Inszenierungen des Gorillas nach 1900." In *Sichtbarkeit und Medium: Austausch, Verknüpfung und Differenz naturwissenschaftlicher und ästhetischer Bildstrategien*, edited by Anja Zimmermann. Hamburg, 2005.

Lechtreck, Hans-Jürgen. "Evolution vor der Kamera: Roger Fenton und Richard Owen im British Museum, 1856–1858." *Fotogeschichte* 109 (2008): 39–56.

Die Linden: Vom kurfürstlichen Reitweg zur hauptstädtischen Allee; Ausstellung der Staatsbibliothek zu Berlin—Preußischer Kulturbesitz zum 350jährigen Jubiläum der Straße Unter den Linden. Berlin, 1997.

Linnenberg, Friedrich. "Eduard Pechuel-Loesche als Naturbeobachter." *Mitteilungen der Fränkischen Geographischen Gesellschaft* 10 (1963): 40–56.

Löschenburg, Winfried. *Unter den Linden: Geschichten einer berühmten Straße*. Berlin, 1891.

Maire, Ernst. *Der Gorilla*. Berlin, 1877.

Meder, Angela. *Gorillas: Ökologie und Verhalten*. Berlin, Heidelberg, and New York, 1993.

Mensch und Tier: Eine paradoxe Beziehung. Ostfildern-Ruit, 2002. Exhibition catalog.

Meyer-Waarden, P. F. *Aus der deutschen Fischerei: Geschichte einer Fischereiorganisation*. Berlin, 1970.

Morgenstern, Frank. "Genealogie der Familie Julius Falkenstein." Unpublished manuscript. Leipzig, 2010.

Morris, Ramona, and Desmond Morris. *Der Mensch schuf sich den Affen*. Munich, Basel, and Vienna, 1968.

Mueller, Volker, et al., eds. *Der beständige Wandel: Darwin und das Entwicklungsdenken*. Neu-Isenburg, 2009.

Nott, John Fortuné. *Wild Animals Photographed and Described*. London, 1886.

Oppermann, Joachim. "Tod und Wiedergeburt: Über das Schicksal einiger Berliner Zootiere." *Bongo* 24 (1994): 51–84.

Partridge, Karen. "In Search of Jenny." Feature, BBC Radio 4, 29 September 2009.

Pechuël-Loesche, Eduard. "Die Tagebücher von Eduard Pechuël-Loesche von seiner Reise an die Loangoküste (24.2.1875–5.5.1876)." Transcription by Donata v. Lindeiner in collaboration with Beatrix Heintze. With support from the Deutsche Forschungsgemeinschaft 1997/1998. urn:nbn:de:hebis:30:3–229806.

———. *Volkskunde von Loango*. Stuttgart, 1907.

Precht, Richard David. *Noahs Erbe: Vom Recht der Tiere und den Grenzen des Menschen*. Hamburg, 1997.

———. "Der schwarze Affe Angst." *Geo* (June 2008).

Rieck, Werner. "Johann von Fischer (1850–1901)." In *Die Geschichte der Herpetologie und Terrarienkunde im deutschsprachigen Raum*, edited by Werner Rieck, Gerhard Hallmann, and Wolfgang Bischoff. Rheinbach, 2001.

Rieke-Müller, Annelore, and Lothar Dittrich. *Der Löwe brüllt nebenan: Die Gründung Zoologischer Gärten im deutschsprachigen Raum, 1833–1869*. Cologne, Weimar, and Vienna, 1998.

Rohlfs, Gerhard. *Kufra: Reise von Tripolis nach der Oase Kufra; Ausgeführt im Auftrage der Afrikanischen Gesellschaft in Deutschland*. Leipzig, 1881.

Savage, Thomas S., and Jeffries Wyman. "Notice of the External Characters and Habits of *Troglodytes gorilla*, a New Species of Orang from the Gaboon River; Osteology of the Same." *Boston Journal of Natural History* 5, no. 4 (1847): 417–43.

Schlawe, Lothar. "Illustrationen nach dem Leben (ndL) aus dem Zoologischen Garten zu Berlin." *Bongo* 23 (1994): 35–62.

Schleich, Carl Ludwig. *Besonnte Vergangenheit: Lebenserinnerungen, 1859–1919*. Berlin, 1977.

Schmidt, Johannes E. S. *Die Französische Domschule und das französische Gymnasium zu Berlin: Schülererinnerungen, 1848–1861*. Edited by Rüdiger R. E. Fock. Hamburg, 2008.

Schmidt, Maximilian. *Handbuch der vergleichenden Pathologie und pathologischen Anatomie der Säugethiere und Vögel*. Vol. 1, pt. 1, *Die Krankheiten der Affen*. Berlin, 1870.

Schneider, Jürg. "Bruder oder Bestie? Die 'Entdeckung' des Gorillas im 19. Jahrhundert." In *Fotofieber: Bilder aus West- und Zentralafrika; Die Reisen von Carl Passavant, 1883–1885*, edited by Jürg Schneider, Ute Röschenthaler, and Bernhard Gardi. Basel, 2005.

Schulze, Andreas. *"Belehrung und Unterhaltung": Brehms Tierleben im Spannungsfeld von Empirie und Fiktion*. Munich, 2009.

Siebenborn, Claus. *Unter den Linden: Galanter Bilderbogen um Berlins berühmte Straße, 1647–1947*. Berlin, 1949.

Soyaux, Herman. *Aus West-Afrika, 1873–1876: Erlebnisse und Beobachtungen*. Zwei Theile. Leipzig, 1879.

Specht, Karl August. *Theologie und Wissenschaft oder Alte und neue Weltanschauung*. 3rd rev. and exp. ed. Gotha, 1878.

The Strange Adventures of Andrew Battell of Leigh, in Angola and the Adjoining Regions: Reprinted from "Purchas His Pilgrimes." Edited by E. G. Ravenstein. London, 1901.

Strehlow, Harro. "Beiträge zur Menschenaffenhaltung im Berliner Aquarium Unter den Linden, I: Der Gorilla (*Gorilla g. gorilla*) 'M'PUNGO.'" *Bongo* 9 (1985): 67–78.

———. "Beiträge zur Menschenaffenhaltung im Berliner Aquarium Unter den Linden, II: Weitere Gorillas (*Gorilla g. gorilla*)." *Bongo* 12 (1987): 105–11.

———. "Beiträge zur Menschenaffenhaltung im Berliner Aquarium Unter den Linden, III: Orang Utans (*Pongo pygmaeus*) und Schimpansen (*Pan troglodytes*)." *Bongo* 14 (1988): 99–104.

———. "Exotik am Prachtboulevard: Das Berliner Aquarium Unter den Linden Ecke Schadowstraße." *Berliner Monatsschrift* 5 (1998): 4–12.

———. "Zur Geschichte des Berliner Aquariums Unter den Linden." *Zoologischer Garten*, n.s., 57 (1987): 26–40.

Ullrich, Jessica, Friedrich Weltzien, and Heike Fuhlbrügge, eds. *Ich, das Tier: Tiere als Persönlichkeiten in der Kulturgeschichte*. Berlin, 2009.

Unter den Linden: Photographien. Essay by Dieter Hildebrandt. Picture notes by Hans-Werner Klünner. Afterword by Jost Hansen. Berlin, 1997.

Vaucaire, Michel. *Gorillajäger: Leben und Abenteuer des Gorillajägers Paul du Chaillu*. Vienna, 1933. Orig. pub. 1930 (in English).

Vierus, Dieter. *Schiffe der Welt: Welt der Postschiffe*. Hamburg, 1995.

Virchow, Rudolf. *Anthropologie, Ethnologie, Urgeschichte: Zur Kranologie Amerikas; Arbeiten aus den Jahren 1871 bis 1894*. Vol. 52 of *Sämtliche Werke*, edited by Christian Andree. Berlin and Vienna, 2002.

———. *Briefe*. Vol. 61, pt. 4 of *Sämtliche Werke*, edited by Christian Andree. Hildesheim, Zurich, and New York, 2007.

Voss, Julia. *Darwins Bilder: Ansichten der Evolutionstheorie, 1837–1874*. Frankfurt am Main, 2007.

Wagnis Westafrika: 150 Jahre C. Woermann; Die Geschichte eines Hamburger Handelshauses, 1837–1987. Hamburg, 1987.

Waldeyer-Hartz, Wilhelm von. *Lebenserinnerungen*. Bonn, 1921.

Waterton, Charles. *Essays on Natural History, Third Series*. London, 1857.

Werner, Anton von. *Erlebnisse und Eindrücke, 1870–1890*. Berlin, 1913.

Wirth, Ingo. *Zur Sektionstätigkeit im Pathologischen Institut der Friedrich-Wilhelms-Universität zu Berlin von 1856 bis 1902: Ein Beitrag zur Virchow-Forschung*. Berlin, 2005.

Yerkes, M. Robert, and Ada W. Yerkes. *The Great Apes: A Study of Anthropoid Life*. New Haven, 1953.

Ziegan, Karl. "Alfred-Brehm-Gedenktafel." *Bongo* 30 (2000): 123–28.

Zöller, Hugo. *Die Deutschen Besitzungen an der westafrikanischen Küste*. Vol. 1, *Das Togoland und die Sklavenküste*. Berlin and Stuttgart, 1885.

CREDITS

Images reprinted by permission of Transit Buchverlag; photo sources follow.

Archiv des Autors: 16, 18, 68
Archives du Muséum d'Histoire Naturelle (1858–61): 11, 13
Alfred Edmund Brehm, *Brehms Thierleben. Zweite umgearbeitete & vermehrte Auflage*, Bd. 1 (1876): 110
Deutsche Kolonialzeitung (1884): 46
Julius Falkenstein, "Die Loango-Expedition: Zweite Abtheilung," in *Die Loango-Expedition* (1879–82): 43, 45, 50, 52, 53, 54, 68
Julius Falkenstein, *Afrikas Westküste: Vom Ogowe bis zum Damara-Land* (1885): 38
Die Gartenlaube (1858, 1866, 1876): 34, 78, 80
Geographisches Institut der Universität Hamburg: 48
Paul Gußfeldt, "Die Loango-Expedition, Erste Abtheilung," in *Die Loango-Expedition* (1879–82): 26, 34
Reichstags-Handbuch, 12. Legislaturperiode (1907): 74
Handbuch der Architektur, Vierter Teil, 6. Halbband (1906): 70, 71
Robert Hartmann, *Der Gorilla: Zoologisch-zootomische Untersuchungen* (1880): 1, 11, 95
Illustrirte Zeitung (1869, 1874, 1876): 30, 31, 42, 66, 72, 77
Kleines Schloss Blankenburg (Harz), Nachlass von Robert Hartmann: 89, 93
Leibniz-Institut für Länderkunde, Archiv für Geographie, *Die Loango-Küste in 72 Original-Photographien (35 Blatt) nebst erläuterndem Text von Dr. Falkenstein* (1876): 27
Leibniz-Institut für Länderkunde, Archiv für Geographie: Nachlass Eduard Pechuël-Loesche, 227/32: 86, 87, 107
Mitteilungen aus dem K. Zoologischen Museum zu Dresden (1877): 23, 24
Museum für Völkerkunde zu Leipzig, *Album der Deutschen Gesellschaft zur Erforschung Aequatorial-Afrikas* (1876): 39, 40
Die Natur, Neue Folge (1876): 76, 85

Proceedings of the Zoological Society of London (1848, 1877): 7, 22
Punch (1861, 1877): 19, 100, 108
H. G. L. Reichenbach, *Die vollständigste Naturgeschichte der Affen* (1863): 3, 17
Transactions of the Zoological Society of London, vol. 5 (1866): 5
Ueber Land und Meer: Allgemeine Illustrirte Zeitung (1876): 83
Westermann's Jahrbuch der Illustrierten Deutschen Monatshefte (1873): 2, 36
Wikipedia (public domain): 6, 10, 14, 15, 20, 21, 29, 58, 60, 62, 63, 64, 68, 72, 104, 112, 113
Christa Winkler: 36
Zeitschrift für Ethnologie (1876): 97